A SCIENTIST AUDITS THE EARTH

Other books by Stuart L. Pimm

Food Webs

The Balance of Nature? Ecological Issues in the Conservation of Species and Communities

The Bird Watcher's Handbook: A Field Guide to the Natural History of European Birds, with P. R. Ehrlich, D. S. Dobkin, and D. Wheye

A SCIENTIST AUDITS THE EARTH

Stuart L. Pimm

Rutgers University Press
New Brunswick, New Jersey

Originally published in hardcover by McGraw-Hill, 2001
First published in paperback by Rutgers University Press, 2004

Library of Congress Cataloging-in-Publication Data

Pimm, Stuart L. (Stuart Leonard)
    [World according to Pimm]
    A scientist audits the Earth / Stuart L. Pimm.
        p.   cm.
    Originally published: World according to Pimm. New York :
    McGraw-Hill, c2001.
    Includes bibliographical references and index.
    ISBN 0-8135-3540-9 (alk. paper)
    1. Environmental degradation.   I. Title.

    GE140.P56 2004
    333.72—dc22   2004046808

A British Cataloging-in-Publication record for this book is available from
the British Library.

Manufactured in the United States of America

For Julia, with my love

Contents

Part Three: The Variety of Life   181

List of Maps

Acknowledgments

I am particularly indebted to Jared Diamond, Paul Ehrlich, and Norman Myers for their support with this book from its earliest stages to its completion, helping in every way from reading its various early drafts, correcting the science, to explaining the mysteries of literary agents. I am grateful for my Pew Fellowship in Conservation and the Environment and to Bryan Norton for nominating me for it. It gave me the time and resources to conceive this book.

This book covers a huge academic territory. While what errors remain reflect my inability to digest every argument, those who read one or more chapters (often repeatedly) prevented many mistakes and greatly helped with my understanding of issues outside my own specialty. My thanks to Robert Askins, Tom Brooks, Gretchen Daily, Kendra Daly, Harold Dregne, Rosemary Gillespie, Grant Harris, Lisa Manne, Jack Moffet, Michael Moulton, Elliott Norse, Daniel Pauly, David Pimentel, Jim Tucker, Jim Sanderson, Robert Powell, Mark Williamson, and Don Worster. More generally, during the writing of this book, I was also extremely fortunate to have received continued guidance from Robert Colwell, John Lawton, Jane Lubchenco, Robert May, Peter Raven, Michael Rosenzweig, Tom Lovejoy, Steven Schneider, and Ed Wilson.

My students at the University of Tennessee and, more recently, at Columbia University read many early versions of this book and let me know its weaknesses in the forthright terms that characterize student evaluations. My former colleagues at the University of Tennessee, Jim Drake, David Etnier, John Gittleman, Sandy Echternacht, Gary McCracken, Susan Reichert, and the late Jim Tanner made my nearly 20 years in Knoxville scientifically as well as personally rewarding.

Bob Pitman, Lisa Balance, and Wayne Perryman's contributions are obvious in the ocean section. Even though I can sling a hammock in the rainforest, I had never been out of sight of land (well, at least on a clear day). I arrived on the *Jordan*, in Panama, at 2 A.M. wondering what on Earth experienced ocean-goers would think of me. Wayne's note on my cabin door explaining what I needed to do, where I would find the head, and other essentials, was the first of many kindnesses. The most recent was sending my favorite photograph of me. I'm visible only as a white safety helmet, in a very small zodiac, next to a large blue whale, taken from 500 feet above in Wayne's helicopter.

In Manaus, Bill Magnusson and I explored the forest by night (where he did not manage to find large fer-de-lances by turning over logs with his bare feet) and snorkeled in rivers with flashlights in plastic bags for electric fish. Maria Alice dos Santos Alves at the State University in Río de Janeiro and her wonderful team have been generous hosts to my research team on our visits to the Atlantic coast rainforest. Walking out of the rainforest, stripping down to shorts, and swimming in the ocean at the field station on Ilha Grande is one of fieldwork's great pleasures.

In Kenya, Tom Brooks managed not to kill me in his four-wheel drive vehicle though my reattaching its essential parts with wire and Scotch tape greatly improved my chances. His not having any food for me for the first three days in Kakamega was an equally unsuccessful attempt on my life. Luke Dollar promised me rice and beans during our fieldwork in Madagascar, but the beans were not forthcoming. The best foie gras on the planet (accompanied by superb and inexpensive French cuisine) is in the capital, Antananarivo, which made the transition to weeks of boiled rice even harder. Rudi van Aarde and his colleagues in South Africa made up for all this by supplying the superb wine for which his country is famous, all the while watching spectacular wildlife. Roger Kitching, my host on two extended trips to Australia, also afforded me the chance to visit spectacular places while living very comfortably. Our more than 20 years of scientific collaborations appear throughout the book. Sonny Bass and I have watched more sunrises from helicopter rides across Everglades National Park than I even thought possible when I worked in Hawai'i; those stories are for another book.

Doug Hembree supplied me with all necessary medical advice prior to my trips, stuck my arm with the requisite vaccinations, prescribed the necessary pills, then restored my health when all these precautions failed. My in-laws, Fred and Raisa Killeffer, have always listened to my traveler's tales with interest and enthusiasm, while providing the food and wine that restored my spirits after strenuous trips. My wife, Julia, accompanied me on many of this book's adventures and has ridden in more helicopters than our friends will ever find credible, given her intolerance for hardships. In recent years, my daughters Stephanie and Shama have been along too. It really is their world—and those of their generation; our children are always the main reason we mustn't make a mess of it.

My passion for good food to accompany fieldwork described above and in the Prologue arose from my losing a third of my weight on a 1968 expedition to Afghanistan (and missing a year of college as a consequence). I thank those who accompanied me on that and all subsequent expeditions around the world for generally keeping me out of harm's way, for their companionship, and for their good humor when things went badly wrong.

A SCIENTIST AUDITS THE EARTH

Prologue

"I won't kill you!" Tom Haupman shouted, as I fastened the awkward shoulder harness in his helicopter. "Thank you," I replied, mustering a brave smile. It wasn't Tom I was worried about; it was our journey. Before landing on Maui's northeast slope, the short flight would penetrate clouds boiling over the Hawaiian Islands' most spectacular mountains. From experience, I knew the first part was low over the gentle sides of Haleakala, the island's large volcano. Then, in an instant, the land would fall thousands of feet below us into the caldera. I knew that I would instinctively grab for a handhold as my pulse raced and my stomach knotted.

It was uncomfortably hot. My Nomex flight suit let in no air and there was no sea breeze that morning. We took off and the 100-knot wind cooled me as I leaned toward the open door, one foot on the helicopter's landing skids. Sugarcane fields stretched *mauka*, the useful Hawaiian term for "toward the mountains." The irrigation canals crisscrossed the fields in neat geometric patterns. Cane is a thirsty crop.

Now the land below was at too high an elevation for cane. *Makai* (the word for seaward), there was a forest of Australian eucalyptus trees, and mauka, the grazed pastures of upland Maui. Ahead, a dark bank of cloud with sunshine-bright edges told me it was going to be wet.

A decade earlier, my student Michael Moulton and I had scoured old census data to find out when lowland Hawai'i had gone to sugar. Planters brought in the first Chinese workers in 1852 and the first Japanese in 1868; the great immigrant surge came two decades later. With efficient irrigation, abundant labor, and a new government that ensured low tariffs for U.S. markets, a third of a million acres were planted with cane within two decades. That acreage, Michael and I quickly realized, comprised almost all of the lowlands.[1]

1

Land too high or too steep for cane now held tree plantations. Long ago this land had been cleared of its native forests. With the heavy rains, the massive soil erosion was obvious even to the land managers of a century ago. They imported trees, selecting those that would grow quickly in the islands' diverse environments. Hawai'i has pine forests, eucalyptus forests, dry savannas, and rainforest—international tree assemblies from dozens of countries and from every continent where trees grow. Very little native forest remained. That was where I was headed.

At last we were over the pali, the steep edge of Haleakala's caldera, with its expected effect on my stomach and pulse rate. To the right stretched the reds, yellows, and purples of the crater floor. At its far end, clouds spilled over the rim and created a patch of surprising and gentle green. I could not quite make out the cabin at Paliku; there would be mules grazing in the adjacent paddock.

In August 10 years earlier there had been no clouds. We had enjoyed 4 days of uninterrupted sunshine in the wet forests of Haleakala's north slopes before hiking over the rim to Paliku. Then it rained. Hours before we all reached the cabin, everything we had with us was soaked. We could see so little that we knew we were close only when the goat control team appeared silently through the mist with their pack mules. By the time we reached the cabin, dinner was on and the showers were hot. Both were heated by a large cast iron stove. We gathered around it to dry ourselves.

Before dinner came silence and grace, which I was delighted to find was said in Hawaiian. The goat team's leaders were ethnically Portuguese, but their temporary helpers were native Hawaiian students. The saying of grace and the hula that followed dinner was for their benefit, not ours. It was the first (and only) time I had experienced the language for its own sake. Sentences were rich in vowels and the glottal stops between them: a bird I had wanted to see before it died out, the 'o'o'a'a is pronounced *oh! oh! ah! ah!*, not *owowarar.*

Goats, introduced in the last century, had once numbered in the thousands. They were eating Haleakala's unique plants to extinction. Removing the goats from the crater had worked. Ten years later, the once-broken mamane trees had recovered. Many were brilliant yellow with thousands of flowers, favorites of nectar-feeding birds. The 'amakihi would be slurping down the rich nectar with their feathery tongues, at least until larger and more aggressive i'iwi came to chase them away.

Tom was carefully checking his GPS for the exact coordinates as we circled above the thick mist. We descended through it and landed at the site we all called Frisbee Meadow. Playing Frisbee was a treat on rare sunny days.

Within minutes, Tom had gone and the silence—and cold—descended. Having barely said "Hello" to my waiting graduate students, I was unpacking my down jacket, head-to-toe Gortex, and Wellington boots and asking for a cup of hot cocoa. The flight was no tourist trip to see the sun rise over the volcano. At 6000 ft above sea level on the windward side of the mountain, Frisbee Meadow could not be more different from Lahaina, the popular destination for the island's aloha-shirted tourists, only 30 mi away. In late January, the weather was always rainy and 50 degrees Fahrenheit. At night, when the clouds sank below the camp, the sky would clear, the temperature would drop, and the water buckets would freeze.

I unpacked the fresh food I'd brought. I would slice the raw ahi—tuna—for sashimi with my pocketknife that evening. Succeeding meals would be something canned with rice, beans, or pasta. And I unpacked the box of paperback novels. At this time of year it could rain an inch an hour for days on end, making fieldwork impossible. A good rain could outlast *War and Peace* and *Anna Karenina* combined. It wasn't raining, so we hiked down to the forest, looking for its unique birds, especially the three we were there to study. In this tiny remnant of Hawaiian forest, red apapane, yellow 'alawahio, and i'iwi were still common. Below and in front I saw an 'akohekohe displaying and, shortly afterward, a parrotbill. Those were two of our subjects. The former was the more common, perhaps numbering a few thousand individuals. Only a few hundred parrotbills remained.

The third was the po'o uli—a bird discovered only 20 years earlier. I peered into the dripping, dark forest expecting one to hop into sight at any minute. I had yet to see one, but bird watchers are an eternally optimistic lot.

I was getting sleepy. Clad head to toe in waterproof gear, I simply lay down in a bank of ferns. Hawai'i has no poisonous snakes, no scorpions, no malaria, and, originally, no mosquitoes. (Well, no *human* malaria. The accidentally introduced mosquitoes spread avian malaria, which is one explanation for why native Hawaiian birds have become extinct since the mosquitoes arrived. At higher elevations where the birds survive, there are no mosquitoes. At lower elevations, I would not have been able to sleep without a heavy coating of deet.)

Every naturalist in Hawai'i has the same dream. We walk in a landscape of sea to mountaintop forests, where 130 kinds of unique Hawaiian birds flourish, some pollinating the beautiful and bizarre lobelia flowers. Along the shore are huge colonies of nesting seabirds. In the dry forests, the williwilli tree is still common, its flowers splash brilliant color across the canopy. The Polynesians have arrived and are beaching their seagoing canoes.

A nightmare follows the dream, one full of disturbing questions about the new century. Most of those birds and many of the plants are gone. How many more species of plants and animals will be lost? What is happening to our fisheries? Will we be eating raw ahi even 30 years from now? How much more of our land will be growing cane and other crops or otherwise under our control? Will some genetically engineered crop stretch from coast to tree line to help sustain the 10 billion of us who will inhabit the planet in the future? What will happen to our forests, our soils, and our wild rivers? Will the parrotbill survive a century from now?

No place begs questions about our human impact more urgently than Hawai'i. One can get a good sense of what we have already done by taking a 20-minute helicopter ride, a half-day drive around any of the islands, or just looking mauka from one's hotel window. There is no escape to a different world. From the summit of Haleakala you see the snow-topped volcanoes of the Big Island (Hawai'i itself) to the south. Beyond—and everywhere else—is the Pacific. The ocean looks, and is, enormous. Even by jet, Hawai'i is a 5-hour flight from anywhere. The islands look small and finite, like our planet does from space.

A Global Natural History

The questions I asked of Hawai'i are ones we must ask of the planet as a whole. And now is the time to ask. Like birthdays, the start of a millennium is a completely arbitrary point. Nonetheless, such occasions demand that we pause to take stock. We do not pause at any other time. An immediate and global natural history is required. This book attempts that feat.

A century ago thoughtful people considered our oceans to be unbounded, the fish stocks within them inexhaustible, the prairies a vast wilderness, and the forests unending and impenetrable. That seems so foolish today.

This book is not about what we should do to minimize our impacts. I will not hector you about having many children, driving a large car, eating meat, turning up the heat in the winter, watering your lawn if you live in California, your politics, or your religion. I will not urge you to become green (or indeed any other color). Nor is this book about whether the lives of human beings will be better in the next century. The impacts I will describe already seriously degrade the lives of huge numbers of people. These impacts will grow in magnitude and geographic extent. I stop short of dwelling on these conditions: This book is not about how the planet retaliates and harms us—even though it does.

I do not supply answers to "How many people can the Earth support?" Joel Cohen, of Rockefeller and Columbia universities, has written a book with this exact same title.[2] There is more than one answer. Indeed, there are hundreds of them. The key word in Cohen's title is *support* not *how many?* The Earth supports some of us in great style and comfort. In other countries, a billion or more people live short, desperate lives in appalling squalor. One cannot answer "how many?" without first specifying the mix of rich and poor. Your views on the matter will undoubtedly reflect the category into which you fit. What is absolutely certain is that humanity's future will be massively different from the past, even if we continue on our present course.

My favorite advertisement for *The Times* has a toga-clad Roman pontificating on the capitol steps, surrounded by superficially attentive spectators, their smiles forced and their many daggers and swords barely visible. "Don't you wish you were better informed?" the caption asks. (With its classical allusion, that ad could be found only in a British paper.) It is not easy to become "better informed" about the present, and the ad hints darkly at the consequences of failure. This book is for those who answer "yes" to the caption's question.

It is not hard to fathom why becoming better informed is difficult. A scientist's work isn't complete until it is published in professional journals. Journals select those who write short, compact, and therefore usually difficult papers for their most demanding and specialized colleagues. The criticism is often harsh, and most manuscripts are rejected, not published. The journals *Science* and *Nature* appear weekly, cover the entire spectrum of science, are the most brutally selective, and, as a consequence, are the most prestigious. Within these two journals can be found the majority of the results I will discuss. One reason I wrote this book was to explain these often complex and important results, to distill them to the simple, memorable numbers that we need to have at our fingertips at the start of this new century.

Big Issues

What are these numbers that we need to have at our fingertips at the turn of this century? The first one sets the stage. At the start of the new century, there are 6 billion of us. This number is rising and may double by midcentury. This number is well known, but the others that are its consequences are not. To explain them, I have divided the book into three parts.

Part 1 involves our use of land and freshwater. Land covers roughly one-third of the Earth's surface but generates 99 percent of our food. Our diverse uses of the land are visible in the areas we convert to agriculture and cities, on which we graze our domestic animals, and from which we harvest timber. An all-encompassing single measure to summarize human impacts is to weigh the green stuff—the amount of material that plants produce each year—then ask how much of it we consume. The answer is "not much"—about 4 percent of the total annual plant production for our food, the food for our domestic animals, and the wood we use for building, paper, cooking, and heating. That answer is misleading. It does not include how much green stuff is wasted while we consume the 4 percent. Add those numbers in, and the total human use of plant production comes to about 40 percent of the global production.

Most of the stuff we use is from the warmer, wetter half of the planet where plants grow best. What remains in dry or cold areas is much less suitable. The warm wet places are where forests can grow and the warmest and wettest are the tropical rainforests. We do not use them sustainably because we are continuously reducing their area. For example, the world's tropical forests are shrinking by 10 percent of their original area per decade.

The drier half of the land surface offers less plant production; it is harder to grow crops there. Paradoxically, we use it for grazing our domestic animals, an indirect and less efficient way of getting food. Drylands are harder to use, provide less food, and are easier to abuse. On more than half their area, wind and water erosion has depleted the fertility of the soils. Grazing animals have changed the vegetation in ways that make it less suitable to them.

The common thinking is that if so much of the world is too dry for extensive use, irrigate it. Make the deserts bloom! Water is another universal currency that we are spending freely. Of the rain that falls over land, the land soaks up two-thirds. We use nearly the same proportion of that (two-fifths) to make plant growth possible. The remaining one-third runs off the land into rivers, mostly in remote places or as floodwater. We consume 60 percent of the accessible runoff each year.

Part 2 of this book considers the oceans. "Yes, the fraction of our consumption on land is large, but surely we have barely touched the oceans!" This is the oft-expressed wish of those who hope for vast, untapped resources to feed the billions of additional people expected in the first few decades of this century. This wish is unlikely to be granted.

About 90 percent of the ocean is a biological desert. We use a third of the oceans' annual production in the remaining 10 percent. The fraction is not increasing. To the contrary, overwhelming evidence shows that we are destroying the oceans' ability to supply even what we take now. We appear to be unable to manage natural resources properly either on land or in the sea. Why are we so incompetent? At the end of Part 2, I offer some answers.

Summarizing human impacts in relation to so much "stuff" gives us succinct statistics, yet it misses significant details. Different kinds of animals and plants are not interchangeable. A grass is excellent fodder for a cow, but a neighboring shrub may poison the animal. Some trees supply excellent lumber; others like the western yew tree are forestry trash. Yet the yew's bark contains trace amounts of a compound called *taxol*, a promising treatment for certain cancers. A slice of ahi tuna destined to become sushi is worth 10,000 times the equal weight of other fish we grind up for fertilizer.

From a human-centered viewpoint, not all stuff is created equal. Life's variety matters. In Part 3 of this book, I consider the variety of life and how fast we are diminishing it. There are probably 10 million kinds of animals and plants. No more than 1 in a million—10—should go extinct per year naturally; that has been the average rate across tens of millions of years of Earth's geological past. Yet in recorded history we have been liquidating animals and plants 100 times faster than the natural rate. The rate is now accelerating to between 1000 and 10,000 times the natural rate. A third of all the kinds of animals and plants could be on a path to extinction by the middle of the century.

These are the big issues, the numbers that are essential for any dialogue about our future. So what about the future? If the past few decades are any guide, the next few will be full of huge surprises. Ozone depletion, AIDS, and global warming have shocked us in the last 30 years. Surprise may be certain, but speculating about the new century's surprises is a dubious enterprise.

I will not even comment on global warming. Its consequences to the greatly changed world in which we now live are what cause concern. Earth has warmed before, never so quickly and perhaps by not much, but certainly never with 6 billion inhabitants. The question is not, What is human impact plus global warming? That is merely the sum of what we know about both. Rather, the question is, How do human impacts interact with global warming? Combined, the effects may be much more than we expect from either acting alone. Scientists are trying to anticipate those possibilities with a keen sense of urgency.[3] But while the physical and chemical changes to the planet are increasingly well understood, their biological consequences, though surely severe, are only now being discovered. So I decided to draw the line at what we know for certain about Earth now, worried—as every informed person must be—by what future changes will bring.

Please Do Not Throw Yourself Out of a Tall Building

I do not list all these facts to recite a depressing litany of how bad things are now and how inevitably grim our future will be. Nor am I telling you that the sky is falling and we cannot get out of the way. News about the environment can be terribly depressing—yet this book is unashamedly optimistic. It is a celebration of our spectacular and fascinating world. I have made no attempt to restrain my joy as I encounter its natural history and its peoples. By the time you read the Epilogue you will know that our world is not doomed, it is not fatally wounded, but neither is it healthy. It needs attention, for without stewardship, its wounds will fester. You go to the doctor when you're sick, not because you enjoy being told you are sick, but because the consequences of ignoring the situation could be severe. You know it is in your best interest to be informed so you can act accordingly. You and the doctor discuss your case, and you consider the options. And we will do that too, in the Epilogue, when we have the facts in hand.

Part One

GREEN FOREST, YELLOW DESERT

I t is Kaua'i, in the Hawaiian islands, 1985. We are driving a winding road early one September morning, along the rim of Waimea canyon. Its bottom is thousands of feet below, in deep shadow, and very much on my mind each time Stanford University population biologist Paul Ehrlich takes a tight bend faster than I think he should. His usually inexhaustible supply of bad jokes had run out. It became question time.

"What percentage of the land's annual plant growth do we consume, directly or indirectly?" he asked. An approximate figure came to me as I remembered my undergraduate lecture notes. How much of the total land surface was desert (a lot of area, but little plant growth), ice (similar area, even less growth), forest and grazing land (some of which we use), and cropland (all of which we use)?

Using Douglas Adams's computer Deep Thought in *Hitchhiker's Guide to the Galaxy*[1] for inspiration, I replied, "The answer is 42 percent." The answer had taken Deep Thought (a computer) millions of years to formulate in reply to a question posed by squabbling philosophers about the meaning of life, universe, everything. It seemed a witty reply, something I rarely managed in response to Paul's jokes, and appropriate too. Plant life supports our life, our universe, our everything.

Paul's first comment was rude, then he explained why: "It has taken us months to get that figure. How did you get your number so quickly?"

The following June, Peter Vitousek, Paul and Anne Ehrlich, and Pamela Matson published the details of their months of effort in the scientific journal *Bioscience*.[2] The work is seminal. It takes stock of our effect on the entire planet, not just a tiny piece of it. It was the first scientific effort to approach these big numbers. It is a brilliant paper, one of the most important of the twentieth century. Yet, no science writer has made it the subject of a popular book. Perhaps with the title "Human Appropriation of the Products of Photosynthesis" this is not surprising. Like many other key scientific papers, it is so technical that even professional ecologists do not read it easily. But had the authors called their paper "Man Eats Planet!" with the subheading "Stanford University Shocker: We've Already Consumed Two-Fifths of It," no respectable scientific publication would have considered it.

The paper has two technical terms: *biomass* and the *production* of biomass. They have the same relation to each other as the amount of money in the bank and the annual interest it generates. *Biomass* is how much living stuff the planet has. *Production* is how much new stuff grows each year—the products of photosynthesis. The Stanford University group was asking, What fraction of the planet's biological interest do humans use each year? Put this way, the question has undeniable importance. If humanity took 100 percent of the interest, there would be none left for other plants and animals, and none for any more of us. Were it more than 100 percent, then we would be living on our biological capital and not just the interest. The capital would decrease each year and so too would the interest. The consequences would be appalling. One cannot think sensibly about our future without understanding what Vitousek and his colleagues had to say.

The Ledger of Biological Accounts

This part of the book expands on the dense text and conclusions from their article. I appoint myself the investment banker of the global, biological accounts. Watch over my shoulder as I check the Stanford group's numbers. For although I guessed roughly the right figures, the numbers—both the total and the details—are too important for me (or you!) to accept on trust. We need to understand how they got the numbers, and from where, and how much confidence we have in their accounting.

The first number to determine is the bottom line of income: How much interest are all the accounts generating? Chapter 1 estimates the total annual plant production from the land. Because the land differs greatly, some of the constituent accounts are larger than others and some generate more interest than others.

In Chapter 2, I start to review the main parts of the Stanford calculation—how we withdraw interest from our biological investments. Each withdrawal in this and the next three chapters will be entered into a spreadsheet. My spreadsheet is not their spreadsheet. I will sometimes agree, sometimes disagree or quibble with them, and sometimes come up with a different way of doing their numbers. All these numbers will have to be assessed and the differences resolved eventually. That is what will happen in Chapter 6.

In reviewing the accounts we visit exotic locations from Afghanistan to what was once called Zululand. Yet, like many adventures, this one starts modestly:—in the kitchen. From cookbooks and food labels, I will calculate how much one person eats per day. Multiply this number by 365 days and 6 billion people. Then add in what cows, sheep, pigs, and the rest of our domestic livestock eat and the wood we consume for heating and building. This gives us the biomass that we consume directly. It comes to about 4 percent of the total annual production. That is a modest withdrawal. Why did the Stanford group estimate a number 10 times larger?

To get the food and fiber that we consume directly, we use and destroy much more plant production. For those of us who prefer eating vegetables so fresh that they don't know they are dead, think how much waste goes on the compost pile after you have shucked an ear of corn and how much of the plant was left behind in your vegetable garden. For those barbarians who get corn from a can, there is another way of looking at this. Use your frequent flyer miles and request a window seat on your next plane trip. What fraction of the landscape below is under human domination, and what fraction is wild?

The exact answer depends on your route and destination. Over Europe, eastern North America, or eastern Asia most of the forest has been cleared. These areas are among the planet's best performing accounts: They generate the most biological interest. It is into these once-forested landscapes that we have inserted our agriculture. We also cut forests, particularly those dominated by conifers, for wood. Across the western United States, the land is a quilt of green forest squares and brown clear-cuts. How we manage these forest accounts is the subject of Chapter 3.

Some flights—like those from Atlanta to Río de Janeiro, or from London to Johannesburg—cross largely intact tropical forests. Chapter 4 sees the biological capital of the Amazon and the Congo forests, which also generate a lot of biological interest, shrinking before our eyes. The view below is often so hazy that we cannot see the individual fires that created the smoke and destroyed the forests.

The London to Johannesburg flight also crosses the Sahara and the Sahel, the shoreline of vegetation at the edge of the sea of desert that forms an intermediate belt of drylands before the humid forests of the Congo. Drylands also dominate North America west of the Mississippi, large parts of South America, central Asia, India, and Australia. Chapter 5 finds that these are large accounts, but they generate low rates of interest. We use these least productive lands in the least efficient way, having our domestic animals eat the plants for us before we eat them. Dryland production is often too sparse to use directly. Our animals must concentrate it for us first.

Impacts on drylands seem minimal from the plane: no fields, few roads, and widely scattered cities. On the ground, the small numbers of domestic animals grazing a wild landscape appear to confirm that impression. Closer inspection rejects it. Our impact causes many drylands to generate less interest than they did in the past. Countries differ in their politics and thus their stewardship of the land. Drylands offer extraordinary examples of national frontiers visible even from space.[3]

The journey ends one evening. As we land, the city's lights sparkle like an enormous jewel spreading to the horizon. Land under concrete generates no biological interest at all.

I list what we have seen, write down the withdrawals, and enter them into the spreadsheet. Chapter 6 sums them to the total of 42 percent of the annual global interest. Forty-two percent and counting, since our numbers increase every year. Can't we make more land create more interest? On our journey, we saw how water turned the large but lackluster dryland account into one paying record levels of interest. Hot, and now wet, plants grow quickly. We remembered a splash of bright green in the dusty yellows below. Irrigation followed a river. Where irrigation stopped, the desert began.

In Chapter 7, I ask how much of the world's freshwater is available for our use? Can't we simply irrigate more land? Humans already take 50 percent of the accessible supply. We take all the water from some rivers: The Colorado River no longer flows to the sea. We completely modify others: The Tennessee River is, from start to finish, a series of artificial lakes. Fifty percent—and counting, that is.

1

Billions of Tons of Green Stuff

How much plant production—green stuff—does the land give forth each year? What simpler question could there be, and what more fundamental one to our future? Ordering "tea, Earl Grey, hot," from the food synthesizer in the wall of the starship *Enterprise* is the stuff of science fiction. Like everything else we consume, tea has to grow first. Knowing how much the planet supplies is a basic question—the bottom line in global biological accounting. It is the number that we seek in this chapter.

What is green stuff? Watermelons are mostly water; spruce needles are not. Grabbing a handful of plants and weighing them will not answer the question. We have to dry the plants first. And yes, you can try this at home. An oven will do fine, though it will smell up the kitchen, and too hot an oven will burn the dry plants. The weight of the dried stuff is the biomass.

Biomass is not production. Why do we need two terms: *biomass* and the *production of biomass?* Biomass and production often go hand in hand: more biomass, more production. Likewise, large bank accounts generate more interest than do small ones. Nonetheless, some accounts earn more interest per dollar invested than do others. In a similar sense, while more green stuff usually means more new green stuff, it need not always.

A forest of spruce trees near the tree line on a mountain may be green enough, for example. The trees, however, may grow slowly. Quite small ones may be hundreds of years old. The opposite can be true, too. A New Zealand pasture may have only short, cropped grass, but the numerous sheep safely grazing on it are quickly converting the grass into lamb chops. Indeed, well-watered, fertilized pastures are among the planet's most productive ecosystems. The pasture is a lot more productive than it looks from the short grass growing there. The sheep are eating the grass as fast as it grows. The spruce

15

trees are just sitting there, hardly able to grow much during the time between the June snow melts and the early autumn frosts.

A lot of green, then, does not always grow much new green, while sometimes a little green does. Scientists cannot simply estimate how much green stuff is out there with a handy-dandy greenness meter and then assume that a constant fraction of it will be new each year. Satellites transmitting images of Earth from space do measure greenness. They are not enough. We have to measure growth directly—on the ground, the hard way.

Forest surrounds my house in the Tennessee woods. Most of the new green stuff grows out of reach overhead. The simple solution is to wait for the leaves and twigs to fall rather than go up to get them. Measuring tree litter is easy: Buy several plastic buckets; punch holes in them to let the water drain; place them in the forest and make sure they will not tip over; empty the buckets periodically and weigh the contents; refuse to answer intrusive questions about your purpose.

Ecologist Roger Kitching and I have done this at Whytham Wood in the cool beech forests outside Oxford, England, and also in the warm humid rainforests of southern Queensland, Australia, where Roger now lives.[1] The major practical problem was to find areas where the containers would not be vandalized or stolen. (No, I do not know why anyone would steal a cheap, plastic bucket with holes in it.) Even in Australia's protected forest reserves vandals roam. One set of buckets was knocked over regularly by an Albert's lyrebird, a rare and spectacular rainforest bird.

In England nearly all the leaves fell in three autumn months. In Australia leaves fell every month, though more in dry months than in wet ones. Even so, the amounts were quite similar: about a kilogram of dry leaves and small twigs per square meter over a span of a year.

For many readers the preceding sentence will seem perfectly ordinary, but not for my wife Julia. She is, in her own words, "metrically incompetent." For her, like most Americans, Britons above a certain age, and some NASA engineers sending probes to Mars, the sentence should read "about 2 pounds . . . per square yard."

Thinking Numerically

Metrical incompetents need not reach for their calculators. It makes little sense to worry about precise conversions from meters to yards and pounds to kilograms. In global planetary accounting, one cannot measure things that precisely anyway. A square meter (m²) is about 20 percent larger than a square yard (yd²); a kilogram (kg) is 10 percent larger than 2 pounds (lb). Few numbers in this book are accurate to plus or minus 10 percent. They ought to be, but they are not. You can understand the math in this book by equating yards and meters, and by multiplying kilograms by 2 to get pounds; then you will be close enough. Better yet, many of the numbers will be in tons, and a metric ton—1000 kg—weighs within 2 percent of a nonmetric ton of 2240 pounds.

The bucket-with-holes estimate of plant production has to be too low. Insects will have eaten some of the leaves before they could fall to the ground. And some of the production will consist of the trees growing larger each year—material that will not fall to the ground until the tree dies. The buckets do not catch falling trees very well. (Nor do they measure how much stuff grows below ground.) Yet sophisticated attempts to measure all a forest's above-the-ground production give averages that are not very different.

One kilogram per square meter is a good benchmark against which to compare what happens elsewhere. The simplest calculation of the planet's total production would be to take this benchmark and multiply it by the total land area. So how much land is there?

By way of answering, let me digress for a couple of paragraphs to discuss facts and whether you should believe them. Listing all the data this book needs is not practical. Yet your inquiring mind demands to know. I certainly hope you will demand more than those who accept the fact-free ranting of America's radio talk show hosts.

In answering how much land there is—and many other questions in this book—the key reference is the data compiled by United Nations (UN) organizations. The UN publishes the data in different formats, including on its Web site, in books replete with page after page of huge tables of numbers, and on computer disks. The most readable form of these numbers appears in a collaborative venture between the World Resources Institute, the UN Environment Programme (UNEP), the UN Development Programme, and the World Bank. Oxford University Press publishes this collaboration as a *World Resources Yearbook* that summarizes UN data on global agriculture, environment, and human population.[2] From these yearbooks come the raw materials for the calculations used in this and other chapters.

Those yearbooks suggest that the land surface (excluding Antarctica) covers 130,984,040 km². This estimate raises a second issue about facts. It is rounded to the nearest 10 km², meaning that we are supposed to believe the first eight numbers of the nine-digit number. Aren't you simply in awe of such precision?

By the time that the next yearbook appeared, the land had shrunk 0.5 million km^2—an area the size of France. Losing France is a remarkable event, unnoticed by *The Economist*, my usually reliable source of world news.

The lesson from all this is that in planetary accounting lots of precision in the numbers may appear to be impressive, but most of the numbers do not mean much. The land area is a genuinely easy number to calculate. Wait until I get to the hard stuff. Repeated experiences with UN data demonstrate its extraordinary carelessness with numbers. Throughout the book I treat UN data with caution and try, whenever possible, to find other sources to cross-check them. This is all the more reason why my telling you how to check the numbers is important.

Not all land supports plant life. Most of Antarctica is covered in ice, year round. Other permanent ice caps are on mountaintops in the Himalayas, the Alps, the Andes, and elsewhere. Take away the ice caps and other cold, inhospitable places where essentially nothing grows, and the area that remains is roughly 130 million km^2. *Roughly* means give or take the size of France.

If the average production is 1 kg of biomass for each square meter, then each square kilometer will produce a million times as much, or 1000 tons. The planet's total production will be 1000 times 130 million tons, or 130 billion tons of biomass.

Millions and billions are now flying around the page fast enough to conjure up the words of the late Carl Sagan. What does 1 million km^2 look like, or 1 million tons? The Appendix to this book is a guide. This number of 130 billion tons is an outrageous calculation! It assumes that a single value of plant growth obtained from plastic buckets in forests typifies places as different as the Sahara Desert, the rainforests of Borneo, Lapland's tundra, and the swamps of Brazil's Pantanal. It is incredible to find that this number is also almost exactly the same answer as that of Peter Vitousek and his colleagues. They estimated the land's total production to be 132 billion metric tons.

We humans are like Goldilocks. We want things to be just right. Most of us live in places that are neither too hot nor too cold, and neither too wet nor too dry. "Just right" means "average." Temperate forests also grow in just right places. They produce more plant growth than dry deserts or cold tundra, but less than steamy rainforests and wetter swamps. One kilogram per square meter is a world average, and most of us live in average places. What about not-so-average places?

Measuring, Classifying, and Mapping the Planet

Our planet is a wonderfully varied; encapsulating that variability into a few numbers is not easy. Describing the planet in detail is an impossible task. This book has roughly 100,000 words. If I wrote just about the land and not the oceans, thus far the words should have covered Europe. They do not. Obviously, there is not going to be a lot of detail.

The trick is to divide the planet into big blocks that are similar. Then one can write about typical forests or typical deserts. In principle, ecologists can repeat the measurements of plant growth across the Earth's deserts, tundra, grasslands, forests, and swamps. It is tedious work to do well, and plastic buckets with holes in them do not work in most places. We need another method.

The first requirement is hard to find: a really good pair of scissors. The second is easier: an ample supply of gullible undergraduates. The latter is obtained by way of signs posted on university notice boards. "Earn money while doing important ecological research," they advise. Recruits sign on for the summer and are quickly shipped to the field, issued the scissors, a rectangular metal frame measuring one square meter, and lots of brown paper bags. The clipping begins.

You now guess that the perfect 1-m squares completely denuded of vegetation that dot the countryside are not the work of aliens or pranksters wanting to be mistaken for them. Rise early and see ecologists at work, clipping away, sitting on short, three-legged campstools. Their overseer will be telling them how much they are learning or, when that fails, that this miserably repetitive task "builds character" or "helps you get into graduate school."

Just one clipping yields an estimate of biomass. It is the easier of the two measurements. Production requires several clippings throughout the year. The idea is this. Start in late winter, when little is growing, and clip. Come back a month later when the plants have begun to grow and clip again in a similar but different place. (The first place has no vegetation left, of course.) On average, there should be more biomass on the second occasion than on the first. The difference measures the previous month's production. Repeat throughout the growing season and add up all the differences to determine the year's total. While the plants grow, we must ensure that our scissors reach them before a cow or a grasshopper does. So the plots are fenced off from anything that might eat them.

Good estimates require a lot of clippings. Where do we start? Clipping an example of each of the 130 million km^2 is impractical. We have to take samples.

There is nothing problematic about the broad patterns of plant growth. Everyone in the servitude of houseplants knows they must be kept warm and wet to grow well. (Talking to them is completely optional.) Warm, wet places produce a lot. Places where it is too cold—in the Arctic or on top of mountains—and where it is too dry—in deserts—do not. We have to take samples from all of these places and carefully compile the results.

Explaining the details of the global patterns of plant growth is not a pretty task, however. Ecologists simply do not agree on how to map plant growth. What precisely do we mean by such common terms as *desert*, *tundra*, *grassland*, *forest*, *rainforest*, or *wetland*? This semantic difficulty causes recurrent problems in our ability to add up the global accounts. How fast we are clearing rainforests, the topic of Chapter 4, depends on what we mean by *rainforest*, for instance. It is not going to help if we dispatch our plant-clipping undergraduates to the far corners of the Earth and we do not know what to call those corners when they get there.

I have a lot of experience with botanists. Each has a preferred classification of Earth's plant communities. Some prefer more than one, depending on the time of day. One botanist classifies the planet into 13 divisions. Another has more than 100, grouped into 30 major divisions and 4 super-categories. One popular scheme has about 20 divisions defined by the combination of annual rainfall and temperature. Most introductory ecology textbooks have maps of the one that the author prefers, and the maps differ from book to book.[3]

Comparing the classifications is an introduction to despair. For some places, it does not matter much which classification one uses. Botanists, like the rest of us, call central Greenland an ice sheet. In contrast, the forests of eastern North America can be all "temperate broadleaf" to one botanist, "subtropical" and "hot continental" to another, a mixture of "warm," "woods," "fields," and "farms" to a third, and three categories of "moist forest" to the fourth. My preferred classification of North America has more than 20 divisions in the east, grouped into three major classes, none of which matches the other sets of three classes. Confused?

The second problem lies in deciding where one division starts and another ends. Even when a forest is a forest (and not a "wood" or a "thicket") and a desert is a desert, where do we draw the line between them? Invent gradations? If this appeals to you—as it does to many botanists—then your classification becomes more complicated and the task of taking samples for each division becomes more daunting.

One scheme produced by Jerry Olson and his associates at Oak Ridge National Laboratory has 12 kinds of forests: coniferous forest, broadleaf temperate forest, mixed forests, tropical forests, shrubs and woods, fields and woods mosaics (or savanna), northern taiga, forest and fields, thorn thickets, coniferous rainforest, mangroves, and wooded tundra.[4] This means we must send the plant clippers off to collect representative samples from each of these 12 kinds of forests—and the other 8 kinds of ecosystems in his classification. That seems to be a modest task.

Jerry Olson's classification also wins the day because his numbers are available from the U.S. Department of Commerce on a computer disk. I will use this classification as a guide throughout. On days when I feel like a botanist I hate it—it does not have enough gradations—but at least I can map areas of "tropical forest" and "thorn thickets" to see precisely where he thinks they are.

Exactly how should we map whatever it is that Olson thinks is a "tropical forest"? The most familiar map is the one designed by Gerhard Kremer, the Flemish cartographer who lived from 1512 to 1594. Mercator (Kremer's latinized name) drew lines of latitude running parallel across the page and lines of longitude running at right angles to them. Any journey that follows a constant compass direction is a straight line on this map. That is fine for following advice such as "Go west, young man." Unfortunately, the map it produces makes Greenland look larger than China (it is a quarter the size). The map makes South America and Africa look small, and they are not. Mercator's projection is not going to work (see Map 1).

Area matters greatly to our arguments: more area, more biological interest. So throughout the book I will use an equal area projection. It squishes the high northern latitudes (but not much grows there) and omits Antarctica (even less grows there). It distorts Australia and New Zealand, a distortion my friends there will not quickly forgive, but it does reproduce the planet's surface so that equal areas on the map equate to equal areas of the classification.

In addition to the areas of permanent ice, there are another 8 million km^2 of the land surface where plants barely grow because it is too cold (and sometimes too dry as well). My second map, Map 2, shows the cold places as an example of the equal area projection that I shall use for all the maps that follow.

How much biological interest do these areas generate? On average, these coldest areas plus another 21 million km^2 of the driest areas generate only about 200 tons of biomass per year for each square kilometer.[5]

Across another 25 million km^2 plants do rather better. The areas are still too dry for trees to grow, except sporadically. Most of this area is grassland. In other places shrubs dominate. These include places around the Mediterranean and similar areas such as South Africa, western Australia, Chile, and California. On average, these areas produce about 500 tons of biomass per year for each square kilometer (a tiny fraction of which goes to produce some of the world's best wines).

Forests occupy about 55 million km^2. Forests vary greatly; Olson is right! They range from savannas with widely scattered trees, to lush tropical rainforests, to cold forests atop mountains and at high latitudes. Exclude the warm, wet rainforests and about 44 million km^2 remain. They produce similar amounts of biomass: about 1000 tons per year per square kilometer, somewhat more in the warmer, wetter areas. While we have few measurements for rainforests, they average more than 2000 tons over 12 million km^2.

Croplands cover over 15 million km^2 and have high production, too—an average of about 1700 tons per year per square kilometer. Wetlands, though they occupy little more than 2 million km^2, yield about 1300 tons per year per square kilometer. There are also a few minor ecosystem types that add small amounts to the total.

To calculate how much each of these areas produces in total, we need only multiply how productive it is by how large it is. For the numbers I have given, the total comes to about 115 billion tons.

The Stanford group used different measurements of plant growth and a different classification of the land. Nonetheless, the two numbers differ by 15 billion tons, about 10 percent. Why did I repeat the calculations? I didn't believe them.

Believing What the Numbers Say

It is nothing personal. I do not believe my estimate any more than I do theirs. Scientists are suspicious, critical nitpickers who are always ready to abuse their friends and colleagues verbally and in print over such differences. That is as it should be. In the arcane, acrimonious debate is a statement of how confident we are about the estimates. Differences in the estimates are to be expected, indeed encouraged. Ecologists calculate different numbers all the time. The key to global accounting is to check numbers by using as many different approaches as possible. When the numbers agree, it is great. When they do not agree, we ask, Why not? In the answer, we hope to find ways of making the numbers more accurate.

Is there some technical fix to this problem? Can't we just do this from a satellite in space? There are indeed detailed estimates of plant production based on finely resolved satellite images of the Earth's surface. These studies employ computer models of how rain and temperature affect plant production. Knowing the rainfall and the temperature, we can make the models predict what the plant growth should be. These predictions, however, still must be calibrated against the on-the-ground clipping measurements.

The point of these satellite-based efforts is that total plant growth will vary from year to year, for some years are wetter or warmer than others. It is essential to grasp the natural year-to-year variability of global plant production. Some years will be much better than others, but we still have to eat in the bad years. El Niño years are particularly disruptive. Taken from the Spanish name for the infant Jesus, unusually warm waters spread eastward across the Pacific, disrupting the anchovy fishery off the Peruvian coast, typically at Christmastime. These events occur every few years and have impacts not only on the oceans but also on the world's tropical forests. The Amazon, for example, is a lot drier in El Niño years, and so global plant production is lower.

These satellite-based estimates range from 120 to 130 billion tons.[6] Although these numbers are quite similar to the ones already presented, you may ask why we are not more certain about them. Ecologists should be embarrassed by not having a good answer—a point I return to in the Epilogue. Yes, we should have better numbers to be properly informed of what we do to the planet, and we should compute them every year. Should these differences cause us to dismiss our estimates of human impact as too uncertain to be worth discussing? No.

Look at the differences in the estimates in the context of the rapidly rising human population. Our population grows by 10 percent in about six years. It hardly makes sense to worry about errors of 10 percent when they will be so quickly superseded by our rapidly changing impacts. The data are not perfect, but they are certainly good enough for developing responses to our impacts.

The simple result of this chapter is that the total biological interest the land produces is about 130 billion tons of biomass per year (give or take a few tens of billions). In deriving this number, we see the Earth as not one investment account but a portfolio of them. Some individual accounts generate more interest than others do. Some accounts are larger than others are. There are the green accounts: warm and wet places covered by forests. There are the yellow accounts: dry places that are deserts and grasslands. And there are white accounts: cold places atop mountains and at high latitudes.

The simple result suggests a simple question: How much of that interest do we withdraw from each account each year? The green, yellow, and white accounts differ greatly in what we can withdraw from them. The white accounts are so small that I will not even bother with them. The green accounts are the most important; I must scrutinize them in detail, and I consider them next.

2

What Earth Does for Us—And What We Do to Earth

I keep my most frequently thumbed reference book, *Joy of Cooking,* in the kitchen.[1] The 15 million or more copies sold confirm that this is an exhaustively and carefully reviewed publication. It contains amazing scientific insights. You should not make mayonnaise when there is a thunderstorm nearby; it commands. I tried and failed, yet always succeed on nice days. Not only that, it contains the first critical numbers that I need for the spreadsheet of the withdrawals from the biological investment accounts.

In the front, *Joy* lists how much food I must consume each day, in calories, and how to convert calories into biomass. How much food the 6 billion of us consume each year is undeniably the most relevant number in the spreadsheet. I will do that first. It is still only one of the numbers we need to calculate our total withdrawals of biological interest from the planetary savings account. We need two other withdrawals. How much do our domestic animals eat? And how much wood do we use? These are three parts of the total sum of what the planet does for us.

The total is an interesting number, but it is only a small part of what we need to know. Nothing we do is 100 percent efficient. It cannot be so. We waste plant production to get the plant parts that we eat. An ear of grain is supported by a stalk; its growth is made possible by the photosynthesis of its leaves and the nutrients taken up by its roots. We eat the grain directly but still must use the plant's leaves and roots to get at what we need. Not only can't we eat the whole plant, we can't use all the land for growing plants. We need places to live and roads to get our crops to market, for example.

This chapter will get these numbers, too—and they are huge by comparison with the amounts of biomass we merely eat. Even so it will not be a complete list. The full accounting of what we have to do to get the food we eat, the food our animals eat, and the wood we consume all involves much larger withdrawals from the global biological interest than our direct consumption. It will take the next three chapters to describe them.

How Much Do We Eat?

According to *Joy*, women from 18 to 35 need 2000 dietary calories per day and "adolescent boys and very active men" need 3000. Take 2500 as the average. You demand that I supply a serious reference for these numbers? Joel Cohen in *How Many People Can the Earth Support?* compiles 14 studies of human consumption.[2] The studies span 30 years and use calculations for many countries. The average of these studies is almost exactly the same as *Joy*. And Cohen's book is no help at all when you need the recipe for beef Wellington.

These numbers are in calories, not biomass. The word *calorie* is a source of anguish to those who really want chocolate cake with Häagen-Dazs ice cream after dinner. A calorie measures how much energy the body obtains by burning food. The dietary calorie has a simple relation to the scientist's calorie. (This is defined as how much energy it takes to raise the temperature of 1 gram of water by 1 degree Centigrade.) The dietary calorie is 1000 calories, actually a kilocalorie. I am sorry to tell you that the cake and ice cream really do have the millions of calories you feared they might.

How, then, do we convert biomass into calories? Look the numbers up in *Joy!* Or, while in the kitchen, read the food labels. I picked up four items that had very little water in them. (I did not want to dry them in the oven to get their dry weights.) I knew that two items would be low in calories—what Americans call an English muffin (an item unknown in England) and a Mexican tortilla. The other two—olive oil from Spain and oil-rich Macadamia nuts from Hawai'i—would have many calories by weight. The oil comes in at 8.8 dietary calories per gram, the nuts at 7.3, the muffin at 2.4, the tortilla at 3.3. The average is roughly 5 dietary calories for each gram of food.

The distinguished Stanford scientists did not use such an unscientific approach. They relied on a compilation of many separate botanical studies by Jerry Olson.[3] We imagine the botanists fighting off savage beasts, crossing raging rivers, and struggling back to their laboratories, handing over plant samples for analysis with their last gasp. The average of these exertions is also 5 dietary calories per gram.

The calculation of human consumption is now easy. Please take out your calculator. Worldwide and per year, we consume 2500 dietary calories per day times 365 days per year times 6 billion people. That equals 5.47 million billion dietary calories. One gram of biomass yields 5 dietary calories, so divide by 5. The answer is 1.1 million billion grams of biomass. There are 1000 g in a kilogram and 1000 kg in a metric ton, so this is 1.1 billion tons. Thank you for your patience. There will be no more such tedious calculations.

Most of the energy we get from our food comes from whole or processed grains: rice, corn, wheat, flour, and pasta. The "everything else" is the fruits, nuts, and the spinach your mom made you eat before you could have pudding. And we are not all vegetarians. Some of these roughly 1 billion tons comes from plants other than grains, some from domestic animals, and some from fish.

How much of these 1 billion tons comes from meat and from fish? The health-conscious among us wish to know! This information is also in *Joy of Cooking* or, since it is heavy to lug around, one of those handy pocket guides that tells you how many calories are in the different foods that we eat each day. Take one with you for a week and keep records. Worldwide, the Food and Agriculture Organization (FAO) of the United Nations (one of the agencies that contributes data to the World Resources Institute *Yearbooks*), calculates that only one-sixth of our diet is from meat.[4]

From your diet records this fraction will seem to be too low. *Beef: it's what's for dinner!* the advertisement brags. In beef-producing states like Texas a figure that small is likely to get you run out of town. (As I wrote this chapter, an American talk show hostess was in Texas court for saying on TV that she would no longer eat hamburgers.[5]) Visit a kitchen in rural Africa or India, and you will come closer in your estimate. Most people are vegetarian by circumstance, not choice. Meat is for the rich, who have to be urged to eat less food for their health's sake.

Since most of what the world eats is plant and not animal food, the *Joy of Cooking* estimate of what we consume ought to match the estimates of how much plant food we grow. It does: We harvest about 1.15 billion tons of grains and other plants for ourselves. So, here is the first of three numbers we need to estimate our human consumption. I have numbered and highlighted this entry because it is the first entry in the spreadsheet that we will need to review in Chapter 6.

1   We eat about 1 billion tons of plant growth each year.

The plant growth that we consume indirectly through domestic animals is the second of the numbers. Vitousek and his colleagues reported a figure of 2 billion tons, a number that I wanted to check.

The FAO also reports that worldwide there are more than 1 billion cattle; a similar number of pigs; nearly 2 billion sheep and goats; fewer horses, mules, donkeys, buffaloes, and camels; and many billions of chickens. For each one of these animals we need the animal equivalent of the numbers in *Joy of Cooking.*

I phoned David Pimentel at Cornell University to ask whether such a book existed, hoping he would have the answer and the good grace not to snicker at my reference to *Joy.* "Morrison," he replied, without missing a beat. A trip to the library at a nearby agricultural college and the book was in hand.[6] Table after table details how many calories pigs and cows and sheep must eat when they weigh 50 kg, 60 kg, and up, whether male or female, whether pregnant or not, and so on. It's a complete diet book for farm animals. Farmers love it because, like *Joy*, this book is an institution, with edition after edition appearing for decades. The copy I borrowed was well thumbed and dog-eared.

Putting the average consumption of a pig next to how many pigs there are, I whipped out my calculator, did a quick multiplication, and got the answer to how much the world's pigs eat. Allow for the fact that most of the world's livestock are small and scrawny, not the prize specimens exhibited by your local 4H (if American) or the county show (if British). I repeated this for cows, sheep, goats, and on down the list, then added all the numbers.[7] The result of this effort is almost exactly the same number as the Stanford group. So now I believe them.

2　Our domestic animals eat about 2 billion tons of plant growth each year.

What about the fish we eat? It is a tiny amount by comparison, as Chapter 9 will elaborate. The global fish catch is about 100 million tons, of which we eat only about two-thirds. The rest is processed as fertilizer or feed for livestock. This total is the wet, fresh weight, which is what the fishery statistics report. The total dry weight of the fish we consume is about 23 million tons. This is a small number compared to the few billions of tons of biomass that we and our animals consume on land. The FAO estimates that worldwide we get less than 1 percent of our food from fish, though 5 percent of our protein.

The third and final number is the plant growth that we do not eat. First, we use trees for construction, for paper, for veneer, and other industrial uses. The FAO tables, country by country and year by year, show that the world's commercial timber harvest totals another 1 billion tons of plant biomass each year. Second, the FAO also reports the wood we burn to cook and to keep warm, again country by country and year by year. That is another 1 billion tons. Adding these two parts gives the total.[8]

3   We consume about 2 billion tons of plant growth each year for wood and fuel.

This first part, industrial use, is likely to be accurate. Government agencies will want to compile lumber mill records to assess the success of their forest-based industries. But that second part, fuel wood, is equal to our total consumption of plants for food. Is such a large number possible?

Our everyday experience in modern kitchens is no preparation for the consumption of plant growth for fuel. The nearest experience we have is finding wood to build a campfire. In the field courses I teach, the undergraduates happily build fires and roast marshmallows—at least on the first night, after which they realize what a chore finding wood can be. Then they return home to mom and her electricity or gas. Most of humanity does not.

As an undergraduate on a university expedition to central Afghanistan, I had collected the scarce dry bushes for our first night's campfire. They burned brightly and were gone in minutes. We had paraffin stoves to cook our food and huddled around them for warmth instead.

The next day I met a boy and his donkey. I had not expected to see anyone so far from the village. Our camp was several kilometers east of it and I had ventured farther away. "Salaam alaikum," I said as he passed. In his reply, I understood only his smile and his thirst. He drank the water I offered and walked on. The load of sticks and sage bushes was enormous, but his donkey was not straining under its weight. For the boy and his family, a meal started with the all-day hunt for fuel across the desolate landscape.

Such scenarios are played out each day in hundreds of millions of homes worldwide. For the world's poor, cooking means first collecting fuel, clearing forests and even small bushes at ever-increasing distances from their homes. The FAO's country-by-country estimates of these efforts are wildly uncertain. Many countries don't even report numbers, so the total of 1 billion tons is likely to be on the low side. In Afghanistan, the boy's family could hardly be expected to care how much biomass it burned, as long as it could find it. The table estimates this country's consumption of fuel and charcoal at 5.683 million m^3. How one could do this is quite beyond me. Are they sure it was not 5.684 million m^3? To get within a factor of 2—somewhere between 4 million m^3 and 8 million m^3—would be an achievement.

Despite these uncertainties, there is a pattern. Africans and Asians burn over 70 percent of the world's total. Thirteen countries burn more than 0.75 m^3 of wood per person per year. Ten of them are in Africa.

A cubic meter is a little hard to grasp. In more accessible units, these numbers mean that, on average, East Africans burn about 2 kg of charcoal per person a day. That is four times the weight of plant biomass that each person eats. And these are averages. People in towns with kerosene and other ways to cook their food will use less, while those in the country will use more.

The countries that burn wood and charcoal are poor. Everyday experiences there make the estimate of 1 billion tons seem entirely plausible. Drive the roads of East Africa. It is never hard to find clusters of white sacks with charcoal inside and small boys attending them. Cook with that charcoal on a small open grill. The estimate of 2 kg per day per person survives an empirical test: Buying charcoal is a daily chore.

The countries that burn wood and crop residues all have rapidly growing populations. An increasing amount of the planet's plant biomass will be going up in smoke.

Biomass Consumption Now—And in the Future

One billion tons for food, two for animal food, and two for wood: The global total of human consumption is 5 billion tons per year, which is about 4 percent of the planet's annual production of terrestrial plant growth that the previous chapter estimated. Although it is not a large percentage, it will grow. As our population doubles, so too will our direct consumption of plant production—if we keep up our present patterns of consumption. We will not keep those patterns, however.

Most of our diet consists of grains, but we also feed those grains to our livestock. Turning grain into meat is a rich country's habit; poorer countries raise mostly range-fed livestock. Europeans and North Americans feed 50 to 85 percent of their grain to livestock. Poor countries are determined to imitate them.

The FAO numbers show that, worldwide, more of the grain goes to livestock now than it did when the Stanford group carried out their study. From 1972 to 1992, Africa tripled the percentage of grain fed to livestock from 5 to 16 percent, and Asia nearly doubled the numbers from 9 to 16 percent.[9] Africa and Asia have a long way to go to match the northern continents. Yet the numbers spotlight our global aspirations and thus anticipate future trends. There will be more people and they will want more meat, and more grain-fed meat.

If we eat more meat, then per person the human impact on plant growth will increase, even if our population size does not. The one-sixth of our diet that is meat consumes twice as much plant growth—indirectly through our animals—as does the remaining five-sixths that we consume directly. Were we to get half our calories from meat, our combined consumption of plant growth would rise from its current 3 billion tons per year to 6.5 billion tons assuming our population remains the same, which it most certainly will not. To predict our future consumption we must allow for a growing population and rising aspirations.

It Is Not What We Consume

The 5 billion tons of direct consumption is misleading in a very important way. We waste far more than we consume. The wheat, rice, and corn that we or our domestic animals eat make up only a fraction of the plant's total growth each year—its roots, leaves, stems, and all. Wood for building is only a fraction of the tree that was felled to provide the finished lumber. So what fraction of a plant is edible to us and to our domestic animals? And how much wood do we waste?

Back to the kitchen for the first of the questions. Grains are our main source of food. Think of how little corn one gets from an ear, how much more biomass is left in the husk, and how even more is in the discarded stems and leaves. Only a small fraction of a corn, rice, or wheat plant is the grain we extract from it. Most of an average plant is inedible waste, and some of what is edible is lost to pests that eat it before we do. Yes, you can do these calculations at home, too: haul in the whole plant and the weeds around it, and then dry and weigh what you eat and what you do not.

No, I have not gone to the kitchen and weighed ears of grain and stems and stalks! Just thinking about it is enough to convince me that for every kilogram of food we harvest, there must be a lot more of the plant that we cannot eat. There is a much simpler way of doing the calculation.

FAO numbers show that the area of agricultural crops is 15 million km^2—a number that is likely to be quite accurate, given its importance to our well-being (And different sources all more or less agree on this number.[10])

To answer how much plant production 15 million km^2 represents we need only multiply the area by the biomass that the land produces each year. In Chapter 1 we estimated the production at 1700 tons per km^2, so across the total area of croplands the total comes to 26 billion tons of biomass.

> **4** Croplands produce 26 billion tons of biomass each year.

We withdraw this amount from the planetary account each year in order to have the roughly 1 billion tons of food that we eat. The remaining 25 billion tons of plant production are waste.

Where do we eat those billion tons? Our homes, our cities, and the roads and railways that connect them do not generate plant life. The planet generates almost no biological interest across an area that covers about 2 million km^2. (The Stanford group's estimate is agreed upon by other sources.)

The mass of humanity along the eastern seaboard of North America, coastal Brazil, western and central Europe, coastal west Africa, the Ganges valley of India, Java, China, Korea, and Japan live in what were once completely forested areas.[11] Few cities grow in the desert; Los Angeles is an example. When they do, these cities often have to steal water from someone else's productive land; again, Los Angeles is a prime example.[12]

Urban areas are generally located in what were once productive farmlands. That is why cities grew up there in the first place. Allow for a moderately high plant production of 1500 tons per square kilometer and the amount of production lost is 3 billion tons.

> **5** Our cities and roads cover an area that would have generated 3 billion tons of biomass each year.

What We Take Is Much More Than What We Get

The first part of this chapter presented three numbers. We eat 1 billion tons of plant biomass each year. Our domesticated animals consume about 2 billion tons, and we consume about 2 billion tons of wood for a total of about 5 billion tons. That is only a small withdrawal: 4 percent of the 132 billion tons of biological interest the land produces each year.

The second part uncovered an important contrast. To grow our crops and have places to live we require 29 billion tons—a much larger withdrawal. What the planet does for us is to provide a few billion tons of food, animal food, and wood. In getting those resources, we necessarily use much more. The answers to such questions define this book's purpose. We change the planet in many different ways as we attend to our needs. As John F.

Kennedy might have put it: "Ask not what the planet does for you, but what you do to the planet."

The contrast—using 29 billion tons to get 1 billion tons of food—is not a complete accounting, for we're still missing major pieces! It is the missing pieces that the next chapters must collect. What else do we do to the planet to get our food, and what do we do to get our animals' food and the wood that we use? What do we do to the planet to have water to drink and irrigation for our crops?

To completely answer the question of what we do to the planet, we must leave the kitchen and journey across the planet to inspect how we are managing the accounts. Where should we start? Forests generate the most interest in the planet's biological portfolio. Let's start there.

3

Not the Forest Primeval

Beside me in the plane sit other frequent flyers, their heads buried in spreadsheets and business plans, worrying about their accounts. I look out the window to see how humanity is managing the forest portfolio.

Fly across eastern America, inland of the Appalachians, on a clear winter's day. Across Illinois, Indiana, and Ohio, only traces of forest remain, scattered black squares above the white sheet of snow-covered farmlands. This was once continuous forest; the few remaining trees testify to an impact that takes nearly 100 percent of the available plant growth. From Atlanta to New York, the ancient Appalachian rocks are crinkled into long, hard, forested ridges. The intervening valleys, with their deeper soils, have farms and few trees.

Fly from Salt Lake City to Alaska along the Rockies. In winter, the land alternates between black and white. The blacks are remaining forest squares, with trees standing above the snow. The whites are clear-cuts, where all trees are removed and only snow-covered bare ground and debris are left. Summer reveals a more varied quilt. Chiseled white mountains interrupt squares of fuzzy dark green forest, squares of brown, the fresh clearings, and the softer, smoother green patches of young, replanted trees.

Across England or Ireland, a different artist has been at work. His palette colors dark hedgerows to delineate irregular, bright green fields. Most of the biological interest is being withdrawn here as well. From Tokyo to London, the flight crosses Siberia. Hours of flying consume forests, flat and featureless from this altitude, interrupted by great wild rivers. We are taking much less of Siberia's biological interest.

A cursory inspection of these forests immediately confirms that we are making large withdrawals from these high-interest-bearing accounts. In many places we might be using almost all of the plant production while few areas are untouched.

Locating the World's Croplands

Our first task in this chapter is to ask for a more detailed accounting to expand on this brief inspection. Where are the world's croplands? And what kinds of forests did we clear to produce them?

A map shows the "where," though it's easier to describe three kinds of "where not." First, agriculture is not found in the northern lands that run across Alaska, Canada, Europe, and Russia. It is not found in the mountains of the Himalayas, the Rockies, and the Andes. Second, it is not found in northern Africa, through Arabia, Iran, and Afghanistan to Mongolia, much of the western United States, and most of Australia. Last, it is mostly absent from the center of South America, the Congo basin, and parts of Southeast Asia (see Map 3).

Like Goldilocks, agriculture selects parts of the planet that are just right. We have cleared the better forests and left the poorer ones alone. Europe and eastern North America are temperate forests, just right for human occupation. Siberia is a boreal forest, too cold for dense inhabitation. Boreal forests cover the higher mountains as well. Agriculture also avoids the wetter forests of the tropics and areas too dry to grow trees.

What was the agricultural land before we cleared it for crops? For the 15 million km^2 that Olson classifies as croplands, my computer can compare each location with a global classification of what the vegetation should be naturally to provide the precise answers.[1]

Croplands take their largest bite out of the more than 11 million km^2 that get enough rain and are warm enough to support forests of one kind or another. Of these 11 million km^2, only 1 million are cut into areas that would have been cool boreal forests, as in Scandinavia and northern Canada. Two million km^2 were once moist tropical forests. The rest were once temperate to subtropical forests like the ones that stretch from New England to Florida.

Another 1 million km^2 of crops are in areas that get less than a quarter of a meter of rain a year. In these areas, only irrigation makes agriculture possible. The remaining 3 million km^2 of croplands are converted from grasslands that get less than half a meter of unpredictable rainfall each year.

Locating the Grazing Lands

Amid the croplands are grazing lands also converted from forests. Examples include the bright green fields of New Zealand and Ireland. They are very productive, and about half of the world's livestock feed on mostly temperate forests converted to grazing lands like these. As such, they are completely human-controlled ecosystems. Just as we do for croplands, we should count all their plant production in our totals of human impacts. About 2 million km^2 of grazing lands are converted forests, making their contribution about 3 billion tons.[2]

> **6** Forests converted to grazing lands account for 3 billion tons of plant production each year.

The other half of the world's livestock feed on natural grazing lands, where they replace or share the land with wild grazers such as deer, reindeer antelope, and wild goats. These conversions of mostly temperate forests to croplands and grazing lands constitute the largest single number in the global spreadsheet. Since they are how we obtained our richest account we should know more about the relevant history.

A Brief History of Trees

This chapter's task is to delve into the details. How were these forests cleared? When were they cleared? Who cleared them? Why did they do so? To be fiscally responsible, we should live on the interest and not deplete the capital. How well are we managing our forest investments?

Only a historical comparison—forests now versus forests past—can fill in these details. That comparison is surely the best indication of what the future of forests will be like. The weary frequent flyers sitting next to me are also projecting their future plans on the basis of past experience. They have their sales figures. I have satellite images from the last 20 years and aerial photography from half a century back. Beyond these, I can delve into old photographs, sketches, and diaries. What lessons about our use of the global forest portfolio will we learn on the fact-finding tour?

My journey to assess the forest portfolio is a long one. It starts with the northern forests, first the temperate ones and then the colder, boreal forests. It is broken at the chapter's end by an overnight flight from Atlanta to the Amazon to visit tropical forests.

Temperate forests lie outside the tropics; most or all of the trees have broad leaves rather than the needles of conifers. Such forests grade continuously into other kinds of forests. These forests, plus those with conifers mixed in (but not ones that are mostly coniferous), and the evergreen eucalyptus-dominated forests of eastern Australia, now cover about 13 million km². This includes the 2 million km² of forests within the mosaic of crops and grazing land[3] (see Map 4).

Where should we ask the questions about these forests' history? In Europe and China, preindustrial societies permanently felled almost all of their forests. These medieval changes are remote in time, so it is difficult to learn the detailed causes. In any case, they may be less relevant to our technology-dependent future than the changes in North America that unfolded within the last two centuries.

There is an added advantage in the search for contemporary lessons from past forest changes in North America: richly detailed accounts. Ecological historians such as Gordon Whitney in *From Coastal Wilderness to Fruited Plain* and William Cronon in *Changes in the Land* describe how, why, and how fast the forest was cleared.[4]

Imagine eastern North America in 1605. What a place the forest must have been before European settlement! Whitney has photographs of the few surviving remnants taken late in the nineteenth century depicting trees that now seem impossibly large. In Illinois the ornithologist Robert Ridgeway stands gun in hand before a 5-m-diameter, 50-m-tall sycamore. In New Hampshire a forester stands among white pines and hemlocks a meter in diameter; another, ax in hand, among red spruces only slightly thinner. Decaying woody debris covers the forest floor. An abundance of epiphytes are on the trees and mosses cover the forest floor. It was Longfellow's "forest primeval . . . bearded with moss, and in garments green."

The "forest primeval"? The trees were several hundred years old in places. A typical area of uncut forest held three to six times the biomass of the forest that would later grow in its place. Yet it was neither untouched nor uninhabited. Hurricanes would transect New England once or twice a century, blowing down trees over large areas. Natural fires burned the forest. People lived there.

Who were these people, and what was their impact? The conventional wisdom is only a few million natives. Others, such as Francis Jennings in *The Invasion of America: Indians, Colonization, and the Cant of Conquest*, argue that this is a western myth.[5] Perhaps 18 million natives lived in what is now the United States and Canada. Their low numbers became tragically true only after contact with European diseases.

Epidemics in the early 1600s caused massive mortality in coastal populations—95 percent is a number that appears often enough. How far the epidemics swept inland is less certain. From what we know of today's epidemics, from sea to shining sea is a possible answer. At least a long way inland seems certain.

Clearing the Forest

After the founding of Jamestown in 1607, America was not settled but resettled. King James thanked "God for His bounty in sending the plague to clear the savages from it."[6] The natives who survived the epidemics and their far more numerous ancestors had an impact on the land. Even with the widely accepted estimate of several million people, the continent was no pristine paradise. They completely cleared areas for their corn. Using stone axes, they first girdled the trees to kill them and then burned them. They set fire to large areas to open the understory and improve the browse for the deer they hunted. These light fires did not kill the larger trees. Different areas burned in different years, concentrating the deer in smaller, more productive burns.

The European colonists cleared the land in much the same way, but in larger numbers and over larger areas. In England, high land values, high food prices, cheap labor, and high densities of farm animals that generated manure encouraged crop rotation. Shortages of capital and labor quickly converted the American colonists to the new ways. Grains were frequently cultivated without manure, which produced only a quarter of the yields in England. Soils were depleted; when yields fell, farmers cleared new fields. The farmers practiced agriculture not so differently from what we see today in the world's tropical forests. Forests would reclaim the old fields, then be cleared again a couple of generations later, and the cycle repeated.[7]

If land had to be cleared, why not clear the best? Few businesses were as profitable as buying and selling land, moving westward, and repeating the cycle. Washington, Jefferson, and other prominent men of that generation were land surveyors. They learned empirically that different forests indicated different soil fertilities. Colonists cleared the best land quickly and completely. They avoided wetter land, for standing water threatened the crops. Besides, malaria was more prevalent there.

Even the uncut forests suffered. Hogs grazed in them, thriving on the abundant supply of beechnuts, acorns, and tubers. After a couple of years of free ranging, they were lured back with corn and packed off to market. Hogs were abundant and fecund, and hence caused a massive decline in beech and oak. Cattle, turned loose in the summer, also retarded the forests' regeneration, turning them into open parklands with a few old trees and pasture.

This destruction was neither simultaneous nor necessarily permanent. An accelerating wave of deforestation spread from the Atlantic Coast to the edge of the prairies to the west, followed by a wave of forest regeneration brought on by farm abandonment. Americans were always moving westward, looking for better land to replace what they had depleted.

Statistics for forest cover in townships and counties in northern states document the patterns.[8] Simply recounting those dry statistics does not do justice to the story. After all, this is the stuff of American legend! And legends require a hero, a fictitious ancestor who could belong to so many American families. Our hero was born in Petersham, Massachusetts, in 1770, on a small farm cleared from the near-continuous forest.

In old age, his world is a very different one from that of his youth. Other authors have written of the many political changes in his life. My tale, however, is an ecological one. By age 70, only 30 percent of the forest around Petersham remains, but increased farming has not brought security. To seek fertile land, his children have had to move westward. His granddaughter, born into the forests of Wayne County, Ohio, in 1810, has a similar experience. On her seventieth birthday in 1880, only 20 percent of the county's forest remain. By then she too has moved westward to Cadiz township, Wisconsin, with her son, our hero's great-grandson.

Not so far away, author Laura Ingalls Wilder writes about her Wisconsin experiences in one of America's favorite children's stories.[9] A little girl lived in the "big woods" of "great dark trees" around "a little gray house made of logs." Beyond those trees "were other trees and beyond them more trees . . . as far as a man could go north in a day, or a week or a whole month." Grandpa shoots the panther that scares him.

On our hero's great-grandson's seventieth birthday, in 1900, only 10 percent of Cadiz forests remains—all but cleared within a lifetime. When the forest went, so did its panthers and wolves. The great-grandson was not there to witness their killing. Both he and Mrs. Wilder had gone even farther west in their relentless search for good land. They are living in little houses on the prairie.

Forest clearing reached its peak in the late nineteenth century, when logging and agricultural clearing were particularly intense in the states around the Great Lakes and in the South. As late as 1850 much of the inland forest was still uncut. Between 1850 and 1909, 22 percent of the eastern forest was destroyed. By 1860 lumber production was the second largest manufacturing industry. (Cotton was king.) Lumbering cleared more land than agriculture after 1880.[10]

The eastern United States once had roughly 2.5 million km² of forest (about a quarter of the current extent of the world's tropical forests). By 1920 only about 4 percent of the forest that had never been cut remained. Maine and Pennsylvania retained less than a tenth of 1 percent of theirs (24 and 57 km²). Ohio had less than 4 percent.[11]

The first lesson from America's ecological history is that large areas of forest were quickly and completely cleared, and this was accomplished by using nineteenth-century technology. Taking the best estimates of changes in forest cover for the region, Robert Askins at Connecticut College and I calculated that roughly half of the area covered by the eastern forest at the time of European settlement (1620) was wooded when it reached its low point in 1872. Almost all of those forests had been cut at one time or another, but some forests had returned.[12]

Since then, forest cover has increased. Although the forests were cut quickly and thoroughly, agriculture did not replace all of them. Forests grew back across more than half the area, particularly on the poor, upland soils.

Coniferous Forests

The world's coniferous forests now cover about 13 million km².[13] Relatively small areas of pine forests exist in warm places such as the southeastern and southwestern United States. Most of the forests are boreal forests that ring the Arctic from northern Europe, across Siberia, across Alaska and Canada, and spread southward along high mountains, including the Himalayas of Asia and the Rocky Mountains of North America (see Map 5).

Where accessible, their fate has followed the deciduous forests to the south. Most of these forests, however, are colder or in mountains and have been harder to exploit. Their exploitation is not history but current affairs. The fragments of uncut forests or old-growth forests of the western United States are the battleground between loggers who want the remaining 5 percent to support their lumber mills for a few more years and those who find other values in the huge trees.

The largest block of boreal forest is Siberia. This is the forgotten forest, its very name a word for a cold, inhospitable place, one of exile and banishment. Its forests survive some of the harshest winter climates on Earth, which is why Siberia still has trees—about 6 million km² of them. That is an area about the size of the Amazon. Two-thirds of this is coniferous forest; the rest is birch forests and mixed deciduous and coniferous trees.

A rough estimate is that 40,000 km^2 are cut each year.[14] This is a rate of less than 1 percent per year, which would allow trees to live 100 years or more before they might be cut again. *Rough* may be too flattering a term for this estimate, however. This is remote Siberia, and statistics from the former Soviet Union were famously unreliable even for accessible areas. At this cutting rate, uncut forests should last until the middle of the twenty-second century, by which time old forests could have grown back where forests had already been cleared. Such optimistic predictions are exceedingly dubious.

Anatoly Shvidenko and Sten Nilsson at the International Institute for Applied Systems Analysis in Laxenburg, Austria, took a closer look at the official statistics from 1966 to 1988, a period before the breakup of the Soviet Union.[15] The area of Siberia's forests grew modestly over that time, according to the official statistics. More telling is how much wood was extracted, how much wood was wasted, and how much less wood there is to harvest each year. The answer seems to be that the harvestable wood declined between 10 and 20 percent during 21 years of study. Rates decline when there is less wood to cut.

Logging is not all that kills trees; there is pollution, of which the former Soviet Union was the grand master. Shvidenko and Nilsson estimate that 65,000 km^2 of forest were damaged by pollution. One cluster of industries at Norilsk, along the Yenesei, alone damaged 20,000 km^2, an area the size of New Jersey. They think this official estimate is "seriously underestimated," which is understandable when you consider that in Soviet environmental writings, pollution and waste were considered to be a symptom of capitalism and thus impossible under communism. Drawing attention to inconvenient facts was not encouraged.

Moncegorsk is not in Siberia but amid the boreal forests in European Russia, north of the Arctic Circle. The smelter here might easily substitute for those found farther east. The low winter sun shines through multicolored clouds of smoke and steam, dimly lighting dead trees over a large area. Stunted ones are farther off. The colors and shapes are reminiscent of a Renaissance painter's vision of Hell.[16]

When I visited there in 1989 the air made me gag and stung my eyes. For tens of kilometers downwind, noxious gases had burned vegetation to the ground. I tried to take photographs discreetly through the bus window, but the driver saw me, stopped, and invited me to step outside for a better view of the mess. My trip took place during the last days of the Soviet Union, but even then, as visiting foreign scientists, we were escorted by tall, uncommunicative men whose presence made our Russian hosts nervous. Our tall man was furious but did not stop me; the KGB's grip was weakening.

Fires destroy the forests, too, and the official estimate is that they destroy 10,000 to 15,000 km² of forest each year. Some foresters regard these estimates as far too low. Even taking the numbers at face value, the fires kill areas almost as large as those harvested. Which of these fires are natural and which are set carelessly is hard to tell.

How Fast Forests Are Cleared

The history of boreal forests echoes the history of deciduous ones. We have cleared them quickly too, whenever possible. The ones that we have not cut have been spared because they are too remote. We cleared forests as fast as we could, but what determines "how fast"? The answers to the question involve technology. Take another look at the history.

In eastern North America, wood was the only source of heating and cooking for centuries. Open fireplaces were inefficient, with perhaps 90 percent of the heat going up the chimney. Gordon Whitney provides numbers that allow a rough estimation of how fast forests would be cleared. Church records report that New England ministers were allocated 30 to 60 cords a year.[17] A *cord* is a 4′ × 4′ × 8′ stack of wood, or about 3.6 m³. The stack is not solid wood. If it weighed a ton, the minister's household would consume 30 to 60 tons of biomass per year.

How much does this consumption compare with the food the household would eat? Suppose that the minister and his wife have eight children. The ten of them would each consume half a kilogram (dry weight) of food per day, or slightly less than 2 tons per year among them.

These are wildly uncertain calculations, but they make a simple point. Without technology or fossil fuels we are not what we eat but what we heat. A bad pun, but at least it reminds us that nothing has changed for the world's poor.

The most impressive old forests in the eastern United States held 300 to 600 tons per hectare. So each household would clear the area of a football field of the best forest every few years just for heating and cooking. A large estate, such as George Washington's at Mount Vernon, used several cords a day. The second-growth forests had much less biomass per hectare, and an even greater area was cleared each year to supply the fuel needs.

The replacement of open fireplaces with the efficient Franklin stove greatly reduced the need for wood. Although these stoves were 4 to 5 times more efficient than an open hearth, they were not in widespread use until about 1820. Coal overtook wood as the main fuel only after 1870.[18]

It was not just the need for fuel that drove the forest changes. Whitney has a long list of less obvious factors that had local, transient, and yet major effects on the forests. Consider the price of beaver skins for hats. When the price was high, the beavers were hunted. Their dams, and the flooded landscapes they created, disappeared. Some trees can benefit—for a while. Maple sugar encouraged settlers to protect the maple trees that produced it. The availability of sugarcane in Louisiana in the early 1800s forced the industry into decline.

Then there was the fluctuating price of potash from wood ash—a fertilizer that sometimes provided a valuable cash crop for the poorer settler and thus another incentive for forest clearing. Leather tanners killed innumerable hemlocks in Pennsylvania for their bark. At the start of the 1800s, less than a tenth of England's pig iron was made with charcoal. Yet in the mid-1800s, three-quarters of the furnaces in the northeast United States were still using charcoal and clearing large areas of forest to obtain it. Some of these factors seem almost quaint, but they all had important consequences to forests sometime, somewhere.

Commercial forestry dominated the later stages of the clearings. At first, foresters selectively cut such valuable trees as white pine and red spruce. Pine harvests peaked in Maine in 1850; the much larger total harvest today is for pulp. In Pennsylvania, the pine harvest peaked before 1870 and hemlocks before 1900. Later, the pulpwood industry clear cut large areas.

Lumbering generated large amounts of waste, or "slashings." These supplied the fuel for some of the continent's worst fires.[19] In 1825, 3300 km^2 of Maine burned. In the fires of 1871, 8000 km^2 of Michigan and 5200 km^2 of Wisconsin went up in smoke. Indeed, in the last quarter of the nineteenth century, fires averaged 2000 km^2 per year in each of Wisconsin and Michigan alone. With the railroads came fires caused by sparks from train fireboxes. These fires often destroyed the "seed trees" left behind for regeneration and their fire-sensitive seedlings.

Earlier I asked: "How fast was the forest cleared?" The underlying explanations come down to beaver skin hats, leather tanners, Franklin stoves, railway lines, sparks from trains, wood-ash fertilizer, and the milling practices of sawmills. A long list of economic and technological factors changed the landscape quickly and dramatically. How can we make sense of them all?

How to Summarize Forest History

Paul Ehrlich and John Holdren (from the University of California at Berkeley) had a simple idea encapsulated in what they call the I = PAT ("eye-pat") equation, one they published in *Science* in 1971.[20] Their idea was that total human impact (I) equals the number of people (P) times the consumption or affluence per person (A) times the technologies (T) used to supply that consumption. Put simply, with more people there is greater impact. Yet, even if the number of people stays the same, they will have greater impact as their affluence increases and technology develops. In the context of forests, the equation offers an important lesson. Forests are cleared faster when more people demand greater affluence and employ more sophisticated technology.

"Eye-pat" summarizes America's ecological history. More people meant less forest, but that was not the whole story. Changing technologies also altered our impact, and usually adversely. Efficient woodstoves and, later, coal, spared the forests. But more mechanized forestry quickly cleared the land and left it vulnerable to fires that trains ignited.

Forestry now satisfies our demand for paper and building materials—signs of our growing affluence. Our houses are many times larger than the one-room log cabins of the early settlers. And the mailbox is stuffed each week with more paper than it would take to print our Massachusetts hero's one book—his family Bible.

Siberia's remoteness has protected its forests.[21] Technology will determine how long they will remain. Improved technology could reduce the enormous waste of trees that the inefficient Soviet system encouraged. Siberia's lumber mills are inefficient; logs have to be cut to a rough size first before they can be milled. There is waste at each cut. Joint ventures among Japanese, United States, and European multinational timber companies may provide the money that improves the technology. The "T" for technology in "eye-pat" can grow smaller, thereby reducing impact. More often, the "T" grows larger. The greatest technological change will be new roads and railroads. They could open up vast areas to clear-cutting that have been too remote until now.

The Return of the Forest

Eastern North America and Europe have similar deciduous forests, similar densities of affluent people, and similar industrial histories. Yet the continents are so different. In Europe, the forests cleared in medieval times are still under crops or grazed. Across more than half of eastern North America forest has returned. What enables some forests to recover while others do not?

America's history was different—and perhaps will be unique—because it had prairies to destroy. Prairies had deeper and richer soils than the shallow, stony soils of the Appalachians in the east. (The more productive bottomlands have remained under cultivation to this day. The forests recovered on the hillsides.)

By the late 1800s the westward-moving Americans realized they could farm these prairies. The Union Army, having nobly enforced the abolition of slavery, found employment for its heroes: ethnically cleansing the prairies of their original tenants. The protected settlers would then write heartwarming stories about their triumphs over ecological adversity. There was a shortage of building materials. Wet prairies nurtured swarms of biting flies and widespread malaria. Dry prairies seemed like deserts to people born in the moist eastern forests.

Once the settlers understood how to drain the wet and farm the dry, the prairies were quickly, completely, and permanently destroyed. In Illinois, for example, less than 3 km^2 of black soil prairies remain from the original total of nearly 70,000 km^2.[22]

Prairie destruction, the most massive ecological change on the North American continent, allowed the eastern forests to recover. After 1920, the amount of deciduous forest steadily grew in the Northeast and the South as trees recolonized the abandoned agricultural landscapes. The size of the trees that surround my Tennessee house tells me that the forests are no more than a century old. To confirm their age, I count the tree rings when they die and fall. In a sketch published in *Harper's Weekly* during the 1870s, the illustration looks southward from the bluff across the Tennessee River to my property. In those days the land was open grazing. So, too, was over 90 percent of this now wooded state.

Return to the westward trek of our fictitious family. By the time they arrived in Wisconsin, the forests in Massachusetts were recovering. Today some of our hero's numerous descendants have returned to live in rural Petersham. They live in big houses in big woods, commute the 80 km to Boston in their BMWs, and the county is again 95 percent forested.[23]

Recover is a relative term. Even in the mountains where the forests have returned, they are often not the same as before. The selective removal of commercially valuable trees contributes to their modern scarcity. Hemlocks, for example, only maintained their presettlement importance in areas that escaped extensive logging. In contrast, weedy trees—those with high reproductive rates, such as birch and aspen—have prospered. Maples and oaks that can sprout from stumps have also done well.

Until I read Whitney's book I had no explanation for the many beeches in the woods around my Tennessee house. Their pale copper leaves, still clinging to the trees on the cold February day that I write this, splash bright color against the black trees and white, snow-covered ground. The beeches are all much smaller than the other trees and have yet to reach the woods' canopy. Farmers once let their hogs forage in the forests where they ate the beech seeds. The practice ended with New Deal programs in the early 1930s that encouraged farmers to pen their hogs.[24] The other trees had a 50-year start on the beeches.

Other changes last century devastated some trees. Any chestnuts roasting on an open fire are now imported from Europe. A fungus, chestnut blight, which was accidentally introduced in nursery stock, eliminated the once-abundant American chestnut trees with surgical precision. At higher elevations in the Appalachians an introduced bug has removed the balsam firs. Only their dead carcasses remain.

Our fingerprints on these eastern forests are clear enough, once you know how to look for them. Very little of the forest is primeval. Only with extended care, decade by decade, will it become closer to Longfellow's image.

What of the coniferous forests in the West? Large-scale commercial logging began during the second half of the nineteenth and early twentieth century and continues to this day. Massive clear-cuts surrounded the isolated volcanic peaks of Mount Rainier and Mount Hood.

Away from the many clear-cuts, the forests may look untouched from an airplane. You can find remote areas of stunning scenery, but this superficial view misleads. Extensive changes proceed below, too subtle to see from this height, at least in most years. If you pick the right year, there is no mistaking the blue-gray haze that obscures the land as plumes of smoke drift downwind. Then you catch sight of the exposed, blackened land.

In 1996, fires in the western United States consumed nearly 25,000 km^2 in one of the worst fire seasons in 30 years. In 2000, it was even worse. The fires likely had a diversity of causes, but one is surely personified by "Smoky the Bear," the cuddly icon who admonishes forest visitors to be careful about preventing fires. That advice is well-meant and sensible, but it is merely the most visible aspect of vigorous fire suppression.[25]

The forests along the humid north Pacific Coast have their famously huge redwoods and sequoias. From Washington state northward, the coastal forests are wet enough for Olson to classify them as "temperate rain forests."

Inland, the forests straddle the half-continent bordered by prairies in the east and the Pacific Coast ranges. Although they are conifers, the forests are much drier, have frequent drought years, and experience regularly occurring fires.

Many of the forests were originally stands of ponderosa pine. The trees were widely spaced, about 1 tree per 200 m², resembling the parkland so attractive to the human eye. Ground fires burned through them every 5 to 15 years. Such fires burn accumulated litter and eliminate more shade-tolerant trees that would otherwise establish under and between the scattered ponderosa pines. In doing so, these frequent fires prevent the buildup of litter that would fuel large fires.

Like the altered forests in the East, settlers also changed the composition of these western forests. In the forests that grew back after clear-cutting, the combination of several decades of fire suppression and the needs of the timber industry has led to a buildup of fire-susceptible and shade-tolerant trees in the understory of the once open pines. These trees, including Douglas-fir, true firs, and lodgepole pine, often reach high densities. One study of northern Arizona forests found that the present-day density of trees is 40 times that of presettlement times.

Such high densities of biomass fuel catastrophic fires. With fewer and smaller breaks horizontally and vertically, fires climb from the ground to the crowns of trees and jump from crown to crown. Large crown fires kill the ponderosa pines over large areas. High tree densities are also far more susceptible to insects and diseases. Worse, pest and fire problems are related and mutually reinforcing. Fungal diseases rot the wood of burned trees. Pest outbreaks kill trees, and, once dead, they are even more likely to burn. So most western forests are not primeval either. Large destructive fires are the most visible proof of our changing them.

Will Siberia's forests recover? Of the 40,000 km² cut each year, 8000 are clear cut with heavy equipment.[26] The machinery destroys most of the undergrowth, postpones tree regeneration, and encourages the growth of low-value weedy trees. Most of the forests are not replanted but allowed to regenerate naturally—if they can. Nearly 65 percent of the cleared areas are on permafrost, areas so cold that the soil is permanently frozen. When its protective forest is removed, the permafrost often melts and becomes a waterlogged bog.

More than 60 percent of the coniferous forests are in mountains. Some areas are both permafrost and mountains. Clear-cuts in mountains often result in massive soil erosion. Even the few areas that are replanted do not do well, for the seedlings are of low quality and are poorly protected.

The answer to the question of whether forests recover depends on what we think is a forest. Do we mean the forest primeval? One with a few tall trees but many small beeches, or dense stands of small pines? How much the forests are changed is another number we must estimate.

How Big Is a Tree?

The eastern United States is wooded, so I ask my ecology classes: "How big is a tree? Not the height, but the average diameter. Open your hands as if to describe the fish you once caught." They stick their arms straight out—a reply that means "half a meter or less."

There are always some who stretch their arms as wide as they will go. They have caught a whopper and probably did so while hiking in North Carolina's Joyce Kilmer Wilderness. This place, one of the few remnant patches of 400-year-old trees, was named after the poet ("I think that I shall never see a poem as lovely as a tree") killed by a sniper in the blasted, tree-less Marne of 1918.

In Kilmer's youth much of the eastern United States was treeless, too. Perhaps that is why he appreciated them so much. Forests have returned to areas of eastern North America that were cleared a century or more ago. The FAO data on forests report that the area of forests in North America and Europe is actually increasing by a modest rate of a few tens of thousands of square kilometers per year.[27]

Just because boreal and temperate forests are not shrinking from year to year does not mean that we are not using them. What the forests' history tells us is that we have converted forests of large, old trees into small, young ones. Temperate and boreal forest may be on the rise, but few living there have seen mature forest. How do we convert this into a withdrawal of biological interest?

I think back to my students and their idea of what a tree is. This, in turn, helps me think about how old a tree is. Yes, I know this sounds like the question, "How long is a ball of string?" But bear with me. To make life simple, I guess that an old tree is now probably 50 years old and an average one half that age. Certainly, there aren't many centenarians! Thus, on average, one-fiftieth of the forested area must be cut, burned, wasted, and otherwise used each year to keep the forests this young. If not, the forests would be aging. (Since the few old-growth forests that remain are going fast, that is highly unlikely.) One-fiftieth of the total area of deciduous and coniferous forests comes to 0.5 million km^2. The consumption of the production of this quite productive forest—say 1500 tons of biomass per square kilometer per year, is three-quarters of a billion tons. The number would be higher if the average age of a tree were less.

> **7** Temperate and boreal forests are harvested, burned, and otherwise used to an amount equivalent to three-quarters of a billion tons each year.

Across large areas of the planet, people might answer my question about the size of a tree by joining thumb to thumb and forefinger to forefinger into a small circle. These are the new forests, the plantations where trees grow fast and die young.

Plantations

Just how much of the Earth's surface is covered by plantations? Plantation trees are a crop, and their productivity must be added to the balance sheet in exactly the same way. Indeed, plantations resemble agriculture with respect to how fast they are harvested. At harvest, eucalyptus trees are cut just above the ground, chipped for pulp, and new trees quickly planted between them. Eucalyptus trees in South Africa are harvested every seven years; Monterey pines, the plantation pine over much of the world, are harvested in a little more than a decade.[28] On the ground, the trees are all young, the same height, and neatly spaced, like corn or wheat.

Some plantations replace original coniferous forests with those of smaller, more tightly packed trees, as in western North America. Conifer plantations now cover growing areas of former upland sheep pastures in Britain—areas cleared of their native forests centuries ago. In some places, plantations have already replaced entire ecosystems. Along these southern coastal plains of the United States, over 99 percent of the once-extensive long-leaf pine savannas are gone, making this ecosystem among the most severely endangered on the continent. Not all plantations replace forest, however. Across KwaZulu-Natal in South Africa and neighboring Swaziland is a large and increasing area of Australian eucalyptus that was once veldt and natural grazing land.

One would expect FAO to be the source of an authoritative, detailed reply to the question of how extensive are tree plantations. It is not. The FAO's report, *Forest Resources Assessment 1990*, covers only developing countries in South America, Africa, and Asia.[29] Together, these total about 0.75 million km². In its sometimes misleading and frustrating way, the

World Resources Institute's *Yearbook* reproduces these numbers without comment, even though they are incomplete.

The *Yearbook* lists no numbers for North America and Europe—all of which have large areas of plantations. There's an old estimate of 130,000 km² for plantations in Europe.[30] There are also huge uncertainties in the estimates for the tree plantations in China. I am not sure why there is not a more confident answer. Whatever the reasons for the lack of information, it is a sorry state of affairs in global accounting, and one that the scientific community must fix. My guess—it is no more than that—is that at least 1 million km² are tree plantations. The number could be nearly twice that.

Tree plantations are among the planet's most productive areas, generating 2000 tons of biomass per square kilometer per year (see Chapter 1). This gives me the next number for my spreadsheet, which I've rounded up to the nearest billion to account for the likely underestimate in the area.

8 Tree plantations account for 3 billion tons of biomass used each year.

What is certain is that this number is increasing rapidly as more of the planet goes under eucalyptus and other fast-growing trees. Viewed from the air, heading south from Atlanta along America's eastern coast, plantations suggest a more modern artist at work. They form oblong patches of uniform color, crisscrossed by white logging roads, a landscape as neat as a painting by Mondrian.

Putting Numbers into the Spreadsheet

As I fly south past Florida's plantations then eastward across the Caribbean, night closes in. I take out the table from the arm of the airplane seat, place my computer on it, open the spreadsheet, and enter this chapter's numbers.

Numbers in the last chapter showed that we burn a billion tons of wood, mostly in the drier countries of the world. It is an efficient use because those who burn the wood likely do not waste much.

We also use a billion tons of wood. To get this 1 billion tons of wood we need the biomass production of 3 billion tons of plantations plus perhaps 0.75 billion tons of more natural boreal and temperate forests that are kept unnaturally young. As for crops in the last chapter, the "what-we-do-to-the-planet" question has a larger answer than the "what-the-planet-does-for-us" question. To get the 1 billion tons we need, we have already accounted for

3.75 billion tons—and that has to be on the low side, for we have yet to visit tropical forests.

The tropics is where we are headed. I so much want to see the Amazon below me, but we will cross it in darkness. Delta's flight affords excellent service, but doesn't Delta understand that the Atlanta to Rio de Janeiro trip could be one of the greatest plane journeys if it were scheduled during the day? Sometime in the middle of the night we fly above Manaus, then continue for hours. Arriving in the early morning, we must catch a flight back along the same path. Apart from Mondays, when there is a direct flight from Miami, this is the only way to get to Manaus. I desperately need coffee. Coffee-exporting Brazil's local airlines serve the most appalling coffee. At least it is daytime, so I'll see the forest and its rivers, and the in-flight music is entirely appropriate.

4

When Vegetation Rioted and Big Trees Were King

Over the headset on the flight northward to Manaus, Brünnhilde is summoning Loge, the fire god, and sending her ravens to torch Valhalla. Grand opera: tragedy among gods and mortals alike, all squabbling over magic gold and the power it brings, the love it denies. It is August, the dry season in the Amazon, the burning season, and Loge has more work here than Wagner ever imagined he could have along the Rhine. Taking off the headset, I smile at the flight attendant, decline another cup of warm water and instant coffee, and say, "I'm sorry, but I don't wish to watch the movie, and no, I prefer to keep the blind open and watch what's happening below." Out comes my laptop to make notes for my diary.

"15th August 1999. 1 PM. For an hour and a half I have been going across a dry landscape with forest only along its few rivers and with huge agricultural fields stretching to the horizon. I just passed a modest town, the largest town for a long while. There won't be another until I reach Manaus, two hours away."

"1:10 PM. The land changes from dry land to forest. As it does, smoke plumes, two large ones, are off to the right. Smoke from half a dozen fires quickly appears, and then a complete to-the-horizon blanket of smoke. I cannot see a thing on the ground. I cannot even count the fires though they must stretch to the horizon. Through occasional thinner areas of haze I see that below is now mostly forest. It forms quite a sharp transition from the fields and grazing land of a few minutes ago. There are some large fields—several square kilometers apiece, I'd guess—cut into the forest. The fires, however, are within the forest itself rather than at its edges. Half a dozen are visible at any one time. Some are huge, sending smoke plumes towering into the sky."

"1:20 PM. We're still in smoke. We must be at the headwaters of the Río Xingu. When I was young I looked at maps and thought this to be the most remote, inaccessible part of the planet, the least known, the least explored. It, too, is going up in smoke."

"1:35 PM. Still mostly forest with large but infrequent farms and a similar number of fires. A few straight roads, but we are a long way from any city. After a few more minutes, we have moved into an area of Pará that has almost completely been cleared of its forest."

"1:51 PM. We have left the uplands and are over the drainage of the Tapajos, another of the Amazon's main tributaries. The land is almost completely forested once again. It is very hazy and I can see little detail. No individual fires are obvious, though only a few create enough haze to make it difficult."

"2:20 PM. We have been over solid forest—rainforest to the horizon—for almost half an hour. We cross a huge river. Must be the Río Tapajos. It is large enough to have huge sandbanks on alternate sides as it meanders to the east. Then we cross a small road running parallel to the river a few kilometers to its north. There is a lot of deforestation along it. By 2:25 PM I cannot see a single human feature below: no roads, no farms, just forest."

"2:36 PM. Another massive river, a dark, black-water river that has flooded the forests far to each side of its main channel. Then, almost immediately, another massive white-water river, this one coffee-colored with sediment and smooth edges. This must be the Madeira and its complex of braided rivers and flooded forest just south of Manaus."

"2:40 PM. A road heads to the southeast. Forest has been cleared along either side of it as far as I can see. Crossroads head bravely into the forest for a few kilometers, and clearings appear along them. This is the road that runs from the south side of the Amazon near Manaus to Port Velho, 700 km to the southeast."

"2:46 PM. A really huge, white-water river. This must be the Amazon. Now an equally large black-water river, the Río Negro, confirms my guess, for they meet at Manaus. Now I see the massive urban blight of the city."

The flight over this southern half of the Amazon took 2 hours—or about 1800 km at the plane's ground speed of 900 km per hour. Half of that time the ground was partly or completely obscured by smoke. Even in the remaining half, fires burned along remote roads.

Although I could not count the individual fires far below me, a satellite did. My student, Clinton Jenkins, downloaded the image for me, and it hangs on my wall. Hundreds of smoke plumes are obvious, some stretch for hundreds of kilometers while others fuse to form a shroud that obscures the forest beneath. Other ecologists would analyze the data and issue their bulletins. In 1997, for example, 24,546 fires broke out in the Amazon basin during a 41-day period in the burning season; compare this with 19,115 fires during 1996.[1] Undoubtedly, the satellite can spot only the larger fires, the ones visible from space.

On landing in Manaus, I left for a field camp that was several hours away by motorized canoe up the Río Negro, the Amazon's largest tributary. I arrived early in the afternoon. The forest around the camp was noisy, but from frogs, not birds. Their strident and varied calls impeded my recognition of the children's laughter, unexpected in so remote a forest. I found the children splashing in the river near a small thatched hut.

Only then did I realize that I had walked across a field of manioc. The plants were far apart, almost choked off by vines and other vegetation, and easy to miss. Gray ash dusted the ground's surface in a few telltale places. These few thousand square meters had been burned probably 1 year, or at most 2 years, earlier. Without the nutrients from this ash, the soil would not yield food; it looked more like beach sand. Looking at the few small yellowish plants, the caboclo would soon be setting another of the many small fires so he could feed his family.

In the last century forests such as these were the final frontier of European exploration. Travelers' tales from them, like Stanley's *Through the Dark Continent*, were best-sellers.[2] They were remote, inaccessible, wonderful, and seemingly inexhaustible. In 1924, the U.S. Department of Agriculture (USDA) *Yearbook* estimated the extent of global forest resources.[3] After detailed state-by-state and country-by-country estimates, often listed to the nearest acre, comes Brazil. A round 1 billion acres—roughly 5 million km²— is arrived at by putting nine zeros in a row across the page and dwarfing every other estimate. Loggers must have thought these forests were inconceivably vast and ageless. In the last century Thomas Huxley thought much the same of the world's fisheries.

Even today, my flight diaries record remote and sparsely populated tropical forests. Later journeys from Manaus to the southwest of Brazil and the homeward journey from Manaus northward to the Caribbean cross even longer stretches with no signs of human impact. Yet we teach our children that these forests, their plants and animals, and the cultures of their indigenous peoples will likely be gone within their lifetimes.

Shrinking Tropical Forests

Is what we teach our children correct? How could these once vast forests disappear so quickly? Could it be that by the time our children have their children the band of vivid green in their school atlas will be recolored dirty orange? How fast are the forests shrinking?

This one journey is enough to remind me that the tropical forest portfolio is shrinking rapidly. I can see the shrinkage, smell the acrid smoke, thrust a hand into the ash that was once a tree. In this chapter I calculate that shrinkage and place it in the spreadsheet of human withdrawals of plant production.

Will the fate of tropical forests be the same as their northern counterparts? We shrank those forests too, centuries ago. The history of humanity in temperate forests shows how long these forests have been our homes and our fields. Will tropical forests become the rich agricultural lands of tomorrow? Are we seeing in the Amazon a repeat of the history that created the wealthy landowners of medieval Europe or nineteenth-century America?

Norman Myers is an independent scientist living in Oxford, England. He has compiled hundreds of scattered references, the analyses of satellite imagery, and his personal experiences into a book published by the U.S. National Academy Press in 1984, *The Primary Source*.[4] He made the first compelling case that tropical forests were going fast.

Myers has an extraordinary track record of being the first to identify and synthesize the truly big issues facing the planet. His work influences more chapters of this book than anyone else's. Synthesis always requires courage; it is easy for other scientists to pick apart one of the many numbers that comprise the total without ever giving a total themselves.

His numbers are controversial, though. At issue is a fundamental distinction in our use of forests. The historical shrinkage of temperate forests has roughly matched the increase in croplands; for one has become the other. Tropical forests are shrinking, so where is the productive land that replaces them? Its absence would damn those who clear the forests as gamblers betting on lousy odds. They gamble not merely with the planet's annual biological interest, but with its capital. (Worse, as Chapter 10 will show, they use taxpayers' money rather than their own.)

The remoteness and huge geographic extent of tropical forests certainly complicate the accounting. Many governments are unwilling to advertise their rapid destruction of their nation's resources, or even let international agencies do it for them. Myers accuses some agency staff of nothing less than cowardice in reporting numbers they know to be politically palatable but scientifically incorrect. I, too, have some choice words about official estimates of deforestation later in this chapter.

We sit drinking the afternoon tea Myers has brewed as I study the photographic evidence on his kitchen walls that displays his admirable efforts to communicate planetary impacts to world leaders. I am as incredulous about his numbers as I am about the other numbers in this book. I confess that I have found compiling, comparing, and contrasting the numbers on deforestation to be among the most difficult tasks in writing this book. As we shall see, Myers's estimates have stood the test of time.

Agreement on forest numbers was not easy to attain. In Myers's words: "After I published my first report in 1980, my hat collected so many arrows from critics that I thought it had more holes than hat. My main finding that the true rate of deforestation was at least twice as large as the former 'official rate' has become part of the mainstream."

The main difficulty in assessing the fate of tropical forests is that these regions are extraordinarily diverse. The now-familiar problem of definitions again creates confusion when one needs to compare estimates. Most people equate "tropical forest" with "rainforest"—a mistake since tropical forests are extremely varied.

Rainforests, defined strictly, receive year-round rainfall and the trees are evergreen—they keep their leaves all year. Such forests constitute only about a third of the tropic's humid forests. Much of the Amazon's forests do not meet this strict definition. Some tropical forests are quite seasonal; varying fractions of trees lose their leaves in the dry season. The forests around Manaus contain such seasonally deciduous trees.

The humid forests themselves constitute only about half of the tropical forests. Dry forests often have too few trees to form the closed canopy of broad-leafed trees that typify the humid forest. Humid forests are mostly tropical but nothing magically happens as one steps across the lines where the tropics of Cancer and Capricorn girdle the Earth. In Australia, large patches of forests 500 km outside the tropics are warm, wet, and evergreen. Their large trees are draped in lianas, covered in orchids and ferns, and have buttressed roots for support. Though the forests are not tropical, birds of paradise call from their canopies.

Nor is there a sharp line as you ascend the mountains. Humid forests stretch to the tree line. They become more temperate—colder and often wetter—as you climb higher and reach the hot cocoa weather described in the Prologue. In the llanos of Colombia and on the fringes of the Congo, tall evergreen forests stretch along rivers, where their green tentacles reach deep into the yellow shades of surrounding grasslands that are too dry for tall trees.

This diversity of forests explains the numbers that follow. The FAO global forest statistics—I am using the publication from 1999—calculate that about 17 million km^2 of all kinds of tropical forests remained in 1990.[5]

About half of this area is dry forests and open woodlands that will be discussed in Chapter 5. Olson estimates about 11 million km^2 for the mid-1980s (see Map 6).[6]

Myers discusses "humid tropical forests" and employs a clear definition. For the "humid" part, he suggests at least a tenth of a meter of rain every month, though one in three years can be drier. They are warm, with a mean annual temperature of 24°C, and hence essentially frost-free. "Wet and warm" means that the trees are at least partly, and often mostly, evergreen. Not every month is equally wet, however. Some trees will drop their leaves in the drier months. The more trees that drop their leaves, the more seasonal is the forest.

Myers estimated an original total for humid tropical forests of about 14 million km^2, less than 8 million of which remained in 1989. Another detailed accounting of tropical forests is in *The Conservation Atlas of Tropical Forests*, published in three volumes, one each for Africa, Asia and the Pacific, and the Americas.[7] Published in the early 1990s, its authors estimate the total remaining forest in the late 1980s to be about 11 million km^2—the same figure as Olson—from an original total extent of about 18 million km^2.

The large differences in the estimates of the original extent of the forests are to be expected. The lines defining their limits are just history. They can differ by millions of square kilometers, depending on where one puts a line through present-day cattle pastures to mark where the long-gone humid forest changed into some other kind of woodland.

Despite all these uncertainties, the simple version of these numbers is that there were between 7 and 11 million km^2 of moist tropical forest in 1990 from an original total of 14 to 18 million. If you want to define such forest strictly, pick a low number; if you accept a broad range of forests, pick a higher one.

Whatever the pros and cons of different estimates, the lesson we teach small children stands. These forests have shrunk by about 7 million km^2 from their original extent. Unlike the clearing of temperate and boreal forests of earlier centuries, these are recent changes. They have shrunk within decades. As I saw during my Amazon flight, they are happening as you read this. So how fast are they shrinking now?

While tropical forests of all kinds are shrinking, it is the humid forests that are now suffering the greatest impacts. Humid tropical forests are in three main blocks: Asia, Africa, and the Americas. They have different histories and are shrinking at different rates. Within continental Asia, large blocks of humid forests remain in Burma, Cambodia, and Vietnam, but they are disappearing fast. The mainland once had over 3 million km² of forest. By 1990 estimates of what remains range from less than 0.25 to 0.33 million km². Myers concludes that the forest shrank by about 300,000 km² over the previous decade.

On the islands of Southeast Asia, including Indonesia, the Philippines, and Papua New Guinea there used to be about 2 million km² of forest. Over a million remained by 1990. The *Conservation Atlas* puts the original numbers at 2.8 million km², which suggests that they include a wider range of forests. They, too, estimate that half of the forest has been cleared. In the preceding decade, the losses were about 200,000 km².

Central Africa still has most of its forests. The original total is 2 to 3 million km². Somewhere between two-thirds and four-fifths remains; the answer depends on whom you ask. The cutting rate in the 1980s was about 65,000 km² every 10 years.

By contrast, in West Africa, the original 1.25 million km² of forest was down to 140,000 km² by 1990. Most of this was in Liberia, Nigeria, and Côte d'Ivoire. Nigeria and Côte d'Ivoire were cutting this forest down at such an astonishing rate that hardly any forest now remains. For instance, during the early 1980s Liberia was cutting only 4600 km² of its remaining 41,000 km² per decade. Nigeria and Côte d'Ivoire, however, were each cutting about 3000 km² per year from their remaining totals of about 40,000 km².

Mexico and Central America once held about 1 million km² of humid tropical forests, half of which was Mexico's. Different sources estimate that between 30 to 40 percent of these forests remain. About 100,000 km² are cleared each decade.

In South America, by far the largest piece of humid forest is in the Amazon and Orinoco basins. The forests covered about 6.5 million km² in 1990, and this represented about 80 percent of the original area. The rates of forest clearing vary enormously, from very little deforestation in Guyana, French Guyana, and Suriname, to extremely high rates in Ecuador.

The smaller piece encompasses the forests along the Atlantic seaboard of Brazil and its southern neighbors. Forests here once covered over 1 million km². Less than 10 percent remains. Myers puts the South American total clearing at less than 50,000 km² per year. By far the largest loss of forest—about 30,000 km² per year—is in Brazil, which contributes the largest single total to the global loss of forests. All these numbers show that what remains of tropical humid forests is going fast.

The important feature of forest numbers is not so much a matter of who has the right ones, but how old they are. An argument over 1 million km² of forest—is or isn't it the right kind?—can last decades as scientists debate. At least 10 percent of the original forest area is cleared per decade—somewhere between 1 and 2 million km², depending on your definition. To update the numbers in the preceding paragraphs, compiled in 1990, one should deduct that area. Yet even this estimate simply extends current rates of deforestation into the future. The problem with the if-current-trends-continue approach is that forest-clearing rates change. Forests shrank faster in the 1980s than in the 1970s.[8]

What Does Deforestation Mean?

If deciding what is *humid forest* is a matter of definition then so, too, is deciding what is meant by *deforestation*. For foresters, it was once a simple matter of noting how much wood was cut, an accounting choice that dominated the reporting up to about 1980. This system now looks pretty silly as vast, cleared areas cover what ought to have still been trees. There is a much longer list of factors.

Myers defines *deforestation* as the complete destruction of forest cover by logging or by clearing for agriculture of whatever sort. He includes cattle ranching, small-scale agriculture, and larger rubber plantations, oil palms, and other trees. He also takes into consideration the large-scale commercial logger, the cattle rancher, and the "shifted cultivator"—farmers permanently squeezed out of their long-established fields elsewhere by more and more people, by violence, or by government actions. By force of circumstance, these shifted cultivators do not allow the forest to recover.[9] (There might be as many as a third of a billion such people worldwide.) However, Myers excludes the shifting cultivator, a different class that includes the caboclo I encountered up the Río Negro, whose activities perforate the forest but allow it to recover. There may be 50 million such people in the world today.

To understand why there is so much more to deforestation than cutting trees, and hence why the rate of forest loss is disputed, let's return to the Amazon. It's the largest single number in the global tally sheet of forest losses.

Amazon Burning

David Skole and Jim Tucker, scientists at Michigan State University and NASA's Goddard Space Flight Center in Maryland, respectively, have done the most comprehensive and detailed analysis of Amazon deforestation.

They compiled the requisite large library of satellite images and, using this collection, published their work in *Science* in 1993.[10]

Remote sensing techniques have greatly advanced our ability to measure changes in tropical forests. There are still problems, however. Deciding where the rainforest ends and other forests begin is still no easier from satellite photos than from the ground. Thick cloud cover sometimes hides the forest below. It does, after all, rain in rainforests! Satellite images are also expensive, and one needs lots of them over many years. And important issues that underlie what we choose to mean by deforestation cannot be detected easily from satellite images.

The largest block of South American rainforest is drained by the Amazon and Orinoco rivers. Most of this forest is in Brazil, but half as much again grows in adjacent parts of Bolivia, Colombia, Ecuador, the Guyanas, Peru, and Venezuela. Skole and Tucker restricted their study to the Brazilian part of the Amazon. This is mostly lowland forest, not the montane forest that laps the Andes. Estimating how fast rainforests are lost from mountains can be a particularly difficult task for satellites. The sun's shadows put some hillsides in deep shade. And in mountains, rain clouds build up in the same places for months on end, continually obscuring the view. The flat Amazon basin itself was easier, though cloud cover prevented about 5 percent of the total area from being measured.

Skole and Tucker first had to define what the Amazon is, and what is and is not moist forest. The Amazon basin of Brazil has a simple legal definition that encompasses six states and parts of three others. Not all of it is forest, though.

Amazon botanists are no more able to agree on what constitutes a certain kind of forest than any other group. Myers counts less than 3 million km^2 of rainforest and adds another half million of "fringe forests." Skole and Tucker count 4.1 million km^2 in moist forests of one kind or another, plus 0.8 million km^2 in dry forest, and 0.1 million km^2 in water. Other scientists chime in with similar numbers.

What all the dueling numbers mean is that there is wet forest, woodlands, and everything in between. I could not see where one ended and the other began as I flew over; everyone has the same problem. Depending on how strict a definition one employs, there is between 3 and 4 million km^2 of forest.

There are big discrepancies in the estimates of how fast this forest is disappearing. For the 1980s, Myers figured that Brazil lost about 30,000 km^2 a year, while the World Resources Institute put the number as high as 80,000 in their 1990–1991 *Yearbook*.[11] Some of these losses would have been in the Atlantic Coast forests, but the great majority would be in the Amazon. A study by the scientists at Brazil's Instituto Nacional de Pesquisas Espaciais estimated only 21,000 km^2 per year.[12] This highest value is four times the smallest one: why the differences?

To witness how David Skole obtains and updates his numbers I visited his lab. A modern building houses the institute and David's lab is overflowing with powerful computers, maps, photographs, digitizers, and undergraduates, postgraduates, and more senior staff working on the images. On receipt, each satellite image is catalogued and the analyses begin.

Everything is computerized, spitting out a large map with boundaries and a first classification of the habitats. The result looks a lot like one of those painting-by-numbers kits: a sheet with lines defining boundaries and numbers inside them: 1 is rainforest, 2 is cleared forest, 3 rivers, and so on. This process saves a lot of work, but computers are still miserable at recognizing patterns that the human eye and brain handle easily. The computer finds it hard to distinguish white blobs of cumulus cloud from areas of cleared forest, for example. So after the computers' attempt, more experienced checkers update the maps against the original satellite photos. In 1978, a little over 78,000 km^2 of forest had been cleared; 10 years later, the total was about 230,000 km^2. This puts the deforestation rate at about 15,000 km^2 per year—only half of Myers's estimate and well below the World Resources Institute's highest reckoning.

The way the forest shrinks is more complicated than our simply clearcutting huge swaths of it. To see why, it helps to visit it. After my visit to Manaus I flew to the southwest of Brazil on the plane that stops at all the provincial capitals. For two hours west of Manaus I glimpsed continuous forest and winding rivers through the streaked gray rain clouds. From Skole's satellite images, I knew this to be one of the planet's largest unbroken stretches of rainforest. I knew also that the scene would change as the plane crossed into Acre and Rondonia—the southwestern states where deforestation is increasing so rapidly. Descending into Río Blanco, thin, bright-green parallel scratch marks ran for tens of kilometers, crossed occasionally by larger perpendicular gashes cut into the dark blue-green of the rainforest. These were the dirt roads and the forest clearing along them that presage complete deforestation.

Skole's satellite images, taken only a few years apart, show first continuous forest, next the giant scratches I saw from the plane, and then the progressive fragmentation of the forest. The pictures are dramatic testimony to the changes that took place during the 1980s, changes that continue today.

Forests are not cleared in a continuous swath. Small blocks of forest remain isolated from the still continuous forest. Such isolated fragments can quickly lose the plants and animals that once lived in them. So Skole and Tucker also measured how much forest was in isolated fragments—the diced and sliced forests. Isolated fragments comprised about 5000 km² in 1978 and 16,000 in 1988. The isolation adds only 1000 km² per year of forests that are severely changed by our actions.

They also measured the area that was near the forest's edge. The scratch marks so visible from both the plane and from space create huge areas that are near forest edges. So what's special about edges?

The Amazon rainforest interior is a surprisingly pleasant place, given its equatorial location. The humidity is high, but the dark forest floor is much cooler than the canopy high above in the sunlight. Walk into a forest clearing; the humidity drops, but the temperature rises, and I, at least, quickly feel uncomfortable. The forest edges are hotter and drier than the interior of the forest. Trees along edges are vulnerable to wind. With shallow roots and no neighbors to protect them, wind can easily topple the first rank of trees, expose the second rank behind them, and so on. Of course, people and their domestic animals enter forests along their edges, as do weedy plants. Edges, like isolated fragments, rapidly lose their original animals and plants (see Map 7).

Skole and Tucker calculated how much forest was within 1 km of an edge. The numbers are dramatic. In 1978, 125,000 km² of forest were within 1 km of an edge. By 1988 the number had risen to 340,000 km². Counting clearing, isolated fragments, and edges together, the rate of damage, 38,000 km² per year, is higher than Myers's estimate.

Skole and Tucker produced detailed maps of these impacts. A crescent of clearing stretches from the Atlantic southeast, then eastward to the Andes, 500 km wide, with fronts of deforestation along the basin's few roads and the Amazon River. That pattern matches the distribution of fires observed by satellite and from my plane journey.

Those maps are now a decade out of date. Has the deforestation changed in that decade? Brazil's Instituto Nacional de Pesquisas Espaciais (INPE) estimate an average of 21,000 km² were lost each year between 1978 to 1988. Since then, INPE has been measuring the forest losses almost annually. The numbers range from a low of about 11,000 km² per year in 1991 to a high of 30,000 km² per year in 1995 (an El Niño year).[13]

These numbers are useful as benchmarks, for they employ a consistent set of definitions, even if they miss many of the processes that cause cryptic deforestation. Even by INPE's conservative estimates the Amazon is going quickly. If the same patterns of forest clearing have been repeated since Skole and Tucker's study—and they surely have—then long before it is gone the surviving forest will be merely a collection of fragments and edges.

This story, remarkable though it is, is not nearly complete. Sitting in the office in front of the computer or flying over the forest in a plane is OK, only as far as it goes. At some stage, you have to go into the field to ensure that classifications are correct and sensible. You have to experience the warm, saturated air; the thunder showers that drench you in minutes; the shots against yellow fever; the nauseating taste of the Lariam pills that ward off malaria; the threat of leishmaniasis; and the ghastliness of its treatment if you contract it.

From the rainforest trenches the complete story is that for every area cleared of forest, sometimes three to four times more forest nearby is severely damaged by selective logging, firewood collection, and especially fires. We still classify such areas as "forest" on satellite images even though the forest is increasingly open.

Even from my plane ride I could tell that forest clearing alone was not the whole story. With few exceptions the fires I could see were within blocks of forest, not at their edges. After the fires died out there would still be forest, of a kind. From my encounter with the caboclo on the Río Negro, I knew that he was puncturing the forest, not clearing it outright.

Estimates of deforestation, such as those generated by the FAO and Myers, at best rely on satellite images. These analyses, however, may significantly underestimate the true extent of the damage from fires and logging.

Daniel Nepstead, of the Woods Hole Research Center in Massachusetts, and a team of Brazilian colleagues detailed the changes in the Amazon basin in a paper in *Nature* in 1999.[14] While satellite images clearly show the areas converted from forest to fields, they do not readily pick up openings in the forest severely damaged by logging. These numerous but small openings soon regrow vegetative cover that can be hard to distinguish from the intact forest. Though logging and fires usually do not kill all a forest's trees, Nepstead and his colleagues demonstrated that they can reduce the forest biomass by nearly half.

By keeping track of selected, well-studied areas, Nepstead and his colleagues found that between 10,000 and 15,000 km^2 were partially logged. That is an area roughly equal to the most conservative estimates of how much forest was cleared by logging or cleared for grazing land. The area that burned during their study was about half as big as the area cleared; so fires are indisputably a contributor to forest changes.

El Niño conditions greatly exacerbate the loss from fires for they bring much less rain to these forests. During the 1997–1998 El Niño drought, fires lit by small-scale farmers burned an estimated 34,000 km^2 of fragmented and natural forest, savanna, regrowth, and farmlands in the northern Amazonian state of Roraima.

The effects of forest clearing, selective logging, and fires are complicated because of how they interact. Logging increases the flammability of the forest because it converts a closed, wetter forest into a more open and hence drier one. That leaves the forests vulnerable to accidental fires from adjacent agricultural lands. Johann Goldammer, from the Global Fire Monitoring Center in Freiburg, Germany, has photographed the rainforest over 20 years, and published a selection in *Science* in 1999.[15] His first photograph shows an intact section of lowland rainforest. Selected logging opened up the forest, and the first fire, in 1982, killed even more trees. Three years after the fire more trees had died, some from drought because the forest had become drier than it was before. After 13 years, even more trees died and the predominant cover was an undergrowth of pioneer trees, which are highly flammable in dry years. A second fire swept through the area in 1998, killing the surviving trees and the pioneers. Soon after, the area was completely converted to a savanna dominated by an invading grass.

Mark Cochrane, also of the Woods Hole Research Center, and his colleagues documented the same process quantitatively, again in a paper in *Science*.[16] After each fire, the forest became more vulnerable to the next fire. Successive fires burned higher into the canopy, spread faster, and burned longer than previous fires. The authors estimated that half of the remaining forest has been affected by at least one fire, thus making a huge area vulnerable to further fires.

I had taken these scientific papers on my latest trip to the Amazon. Until I read them I had not appreciated the extent to which forest fires are set in the forest itself rather than at the forest edge as a means to expand clearings. The fires I saw were not the evidence of forest clearing but the weakening of the forests that would make them vulnerable not only to deliberate clearing but to accidental fires.

The lesson from the Amazon is one of large-scale deforestation, but even larger-scale fragmentation, and larger-scale severe damage by fire and logging. Is this lesson from the Amazon typical?

The Forests of Southeast Asia

On the far side of the world are the forests on the islands of Southeast Asia. Myers's numbers suggest that these forests may last the longest. Yet this region also includes some of the most completely deforested tropical areas, places where hardly a tree remains. It is also a region of extraordinary biological and cultural diversity.

For the Philippines, Myers estimated that only 20 percent of the rainforests remain. This is close to my own estimate of 24 percent of all forest types remaining.[17] With a rate of clearing of nearly 3000 km^2 a year, any difference is quickly swallowed by time. My own estimates agree with those of Myers that for the major parts of Indonesia (the Greater Sundas and Sulawesi), only a little over half of the total forests have been retained. For the country of Indonesia as a whole, reports by the World Bank and the FAO estimated the forest losses at between 6000 and 12,000 km^2 per year.[18]

Just as in the United States of the last century, the Indonesian deforestation creates slash piles that fuel massive fires. In September 1997, bush fires spread smoke across much of Southeast Asia. The smog raised health alarms throughout Malaysia and in Singapore, where residents were advised to minimize their time spent outdoors.[19] At that time, I was visiting Jim Tucker's lab. His computer screen showed a gray, blurry satellite image across this huge area, one made more dramatic by the crystal sharp pictures outside the smoke. The fires burned over 7000 km^2 of Borneo and Sumatra, and were set mostly by farmers and plantation companies to clear land already logged for agriculture. The fires spread to fragments of uncut forest, especially the drier forest on steeper slopes.

The press releases at the time complained that the fires were the worst ever, though a previous one in eastern Kalimantan—the southern, Indonesian part of the island of Borneo—burned 36,000 km^2 in 1983. Those fires ignited the peat underlying the tropical peat swamp forest that is particularly extensive in this region. A decade later some of those fires were still burning, and the 1997 season was likely to ignite more.

At the rate of 12,000 km^2 per year—the best estimate but one that excludes fires—Indonesia is losing forests faster than any other country except Brazil. At this rate, no rainforest will remain in Asia by the middle of this century. Some large islands already have little forest left; Java has less than 10 percent and the smaller islands to the east have only about 15 percent. In short, the story here is the same as in the Amazon. What we take for timber is a small fraction of our total impact. We carelessly burn huge areas of old-growth forest, and the fires readily burn the secondary forest over and over again.

Are authorities trying to protect at least some of these forests? As Myers points out, the areas designated as conservation and protected forests have often been cleared. Protection "on paper" is an unreliable guide as to how large an area of forest will remain in the future. "Forest" on paper may be a complete absurdity.

Most of the Philippines is deforested, and the island of Cebu has the least forest of all. All but a few of the island's 5000 km^2 had been cut by the mid-1980s; Cebu's low elevations are an easier target than the more mountainous islands of the Philippines. Nonetheless, the *Conservation Atlas* lists a large protected area on the island.

Tom Brooks and his Philippine colleague Perla Magsalay set out before dawn from Cebu City to look at what remained. The city had grown quickly to be the country's second largest. Cebu was a model for the developing world according to a gushing article in *The Economist*.[20] To Tom that meant it had MacDonalds and Pizza Huts, street vendors, and jeepneys—ancient minibuses that had spent their better years in Japan and which were now spewing out clouds of black fumes into the clogged streets. "And thousands of school kids, all in uniforms."

He headed for the Central Cebu National Park 18 km away, armed with the government's information sheet. It told him that the Park had been established on September 15, 1937. It said that the park's area was exactly 11,893.5833 hectares—almost 119 km^2 measured to the nearest square meter. "Portions of the area are farmed," the report continued, and there is "occasional cutting of trees for firewood and construction purposes."

"It must have been more than occasional," Tom thought. Farther out, the city gave way not to forest but to intensively farmed hillsides. There were rice paddies on terraces and crops on even the steepest hillsides— except for the frequent scars where erosion had washed them away. Tabunan was a village in the center of the park. Here there were only "hundreds of kids," few with uniforms. They pointed out Tom and his group as curiosities to their friends.

Curiosities indeed. Who would look for forest here? On the steepest slope, Tom found the small patch he had heard about. Surrounded by hill rice and tangles of thick weeds were the remains of Cebu's forest. Clambering up the slope in the heat and humidity, Tom didn't take more than an hour or two to explore all of it.

The Forests of Central Africa

Central Africa holds the third largest of the big rainforests. Over half of that forest is in the Democratic Republic of the Congo (formerly Zaire), an area of over 1 million km^2 with two-thirds of the original forest remaining.

Woody Meltzer described to me his trip there as we sat in his laboratory in Pretoria, South Africa. I decided I wasn't in any hurry to go there and admired him for searching for Menge's killer. Menge had lived in Kikwit, an inland town 500 km south of the Equator on the Kwilu River, a tributary of the Congo. Menge died from the Ebola virus. All those close to Menge had died too. That was simple enough to understand: Ebola fever quickly killed half the people who contracted it because it is highly infectious and untreatable. But why was Menge the first to die? Menge and the others were among the world's poorest people, yet the puzzle of their deaths was of pressing, international concern.

Like the others who lived in Kikwit, those who died started their days before dawn with a 20-km walk to find forest remnants. Sometimes Menge would spend the night in the forest. "By dawn everyone was walking out of town," Meltzer told me. "Women were off collecting the large leaves to wrap food for sale in the markets. Others would go to remote fields burned into the forest to grow maize and cassava. Men carried pangas, some to cut rough planks from trees. By evening they would walk home, toting half a dozen 3-m planks."

Adults and children alike looked for anything that moved: bats caught at night in nets, rats and mice in snares, monkeys if they were lucky, and caterpillars and beetle grubs. All were for sale in the huge local market whose stinking floor was matted with leaves and rubbish. Or perhaps they wanted what Woody thought was smoked monkey to exchange for the fish that the Portuguese trader brought upstream from Kinshasa.

Menge would have eaten all of these delicacies; he also had eaten a snake a week or so before he died. It might have been what he ate or it might have been something that bit him. There were a lot of possibilities. Whatever it was, Meltzer wanted to capture it. "Goats and cattle were unlikely," he told me. "The people were too poor to own them." His task was difficult because the bats, rats, lizards, and snakes ended up in the pot so fast that he couldn't trap them first.

Wherever he went there were children. They made it hard to leave traps and nets unattended, let alone put them out in the first place. The children were still there when he was arrested at gunpoint; they gleefully pointed their fingers to their heads, suggesting an imminent execution. It took several hours before a literate guide read his permit to be in Kikwit and, more important, bribed the soldiers.

Woody's two most significant impressions were the children, "thousands of them," and the difficulty of finding the forest remnants. Those remnants would be home to the unknown animal that provided the reservoir of the Ebola virus that killed Menge and the others. Until Woody and his colleagues could find the Ebola host and teach people how to avoid it, the threat of other outbreaks would remain.

Logging in tropical forests sometimes has unexpected and fatal consequences far beyond what one expects from the loss of plant biomass. Logging camps pay hunters to provide food for the camps, and bush meat is a major source of income and food. Chimpanzees and monkeys are prized items. All are dispatched with predictably bloody methods. Three strains of HIV-1, the virus that causes AIDS, probably infected humans through three separate bloody contacts with chimp bush meat. Also various strains of HIV-2 came from SIVsm, a virus found in the sooty mangabey monkeys of West Africa. Another virus, HTLV-1, appears to have come—and perhaps more than once—from the STLV-1 virus of chimps. And human pygmies living in Gabon have been infected with a strain of HTLV-1 quite similar to the STLV-1 of the mandrills living in the forest. Monkey and ape populations in West and Central Africa are a reservoir for repeated infections.[21]

Imagine a journey eastward from Kikwit, 1000 km across the Congo basin's rainforest. For 100 km, only the smallest scraps of forest remain. Then we skirt the rapidly eroding forest and the unbroken canopy stretches far to the north. Another 700 km later we are still in the lowlands, about to cross the Lualaba River. We turn northeast. Soon the land rises steeply to the first of two giant cracks in the African landmass: the Albertine rift valley.

Over tens of millions of years Africa has slowly split along a westerly curved line from Ethiopia's Red Sea coast to southern Mozambique. Within the split are the long, thin, rift lakes: Albert, Tanganyika, and Malawi. Flanking it are high volcanoes where lava pours from the cracks. Margherita in the Ruwenzoris is Africa's second largest peak at over 5000 m above sea level. To the south are the lower but impressive Virungas, towering into the mists and home to the gorillas that primatologist Diane Fossey made so famous. Toward the northeast, across Uganda and Lake Victoria, the country is lower and drier. Then we reach the smaller, more easterly crack in the continent—the one that runs through Kenya—and its flanking volcanoes. The higher land running southward from Mount Elgon catches enough moisture to create an island of rainforest. Although isolated from the Congo basin and the Albertine rift, the forest is similar.

This hypothetical trip across tropical forests is neither practical nor safe. Eastern Zaire, Rwanda, and Burundi are names that evoke sickening stories of genocide. Kenya, however, boasts a much-respected wildlife service, a large ecotourism industry, and famous national parks. Kakamega, a small fragment of lowland wet forest, even if off the main tourist trails, is well known among the select ecotourists who come to watch its wildlife. It is as close to the Congo basin as I have yet dared to go.

Children stand by rows of white sacks of charcoal. How else would the people here cook? Loaded with huge tropical trees, a logging truck goes by, bearing the name of the company owned by the president's son. There must be forest here somewhere, but it is not obvious where. This is not like Europe, or America, or Australia, with remnant forest fragments scattered among farms. Here, every last native tree has been cut.

Amid the eucalyptus trees, guava, and lantana bushes—the inedible weeds that have inherited all the tropical Earth—more children tend thin cattle. Kenya's population has doubled in the last 17 years, and so half of the populace must be under 17. Mathematics explains what I see, but it does not lessen my surprise.

The forest boundary is sharp, and as we enter the reserve, huge, noisy hornbills fly overhead to greet us. Our young, enthusiastic guide, with a broad infectious smile, shows us around the tourist camp (thatched huts, with neither water nor electricity), the forest, and its birds. Along the trails, small saplings have been sliced off with machetes and women walk back to their village with bundles of them on their heads. Most of the large trees are gone and all the valuable ones have disappeared. One large, decaying specimen has only a small section cut from it, just large enough for a group of men to carry off. Cowbells tinkle; a small boy passes with his cattle. The local member of parliament gained grazing rights in the reserve in exchange for his allegiance to the president's political party. The trails are now cattle highways; lantana and other weeds are moving in.

The reserve does not allow charcoal burners, but our guide had stumbled on some in a rarely walked section of the forest. The forest guards whom he summoned were mostly concerned about selling the charcoal themselves. He called the wildlife service, and the burners were arrested. Judging by how much charcoal is for sale, my impression was that such arrests occur only rarely.

Back in Nairobi I peered at satellite images and older aerial photos of Kakamega taken over the last 40 years. The forest contrasted sharply with the fields; it shrinks and becomes more fragmented from image to image over time. In recent images only a small island remains, now remote from Mount Elgon's montane forest to which Kakamega had once been joined.

The Making of Bad Land

One final question about tropical forests must be answered. Humid forests have shrunk by 7 million km², but only 2 million km² of croplands are in former humid forests. What has happened to the rest?[22]

Most tropical forest cutting does not produce cropland. So is it grazing land? To call it such is to look at the world with more optimism than even Voltaire's Dr. Pangloss could ever muster. The abject poverty and bloody conflict that follow the loss of forest is ample testimony to the fact that most has become simply bad land.[23]

Clearing Europe's forests produced farmland. In the south of England, the land was rich enough that 900 years ago its owners were able to educate their sons at Oxford and Cambridge. Will former rainforests grow a new generation of prosperous farmers to richly endow Rondonia University or Central Congo State? Those who cut the tropical forest do not become the rich farmers of tomorrow. They abandon that land with the same desperate inevitability as the settlers in the Appalachians did two centuries ago.

Will North America be the model of what happens in tropical forests? It now has far more forests than it did a century ago, forests that only a careful observer might easily distinguish from the original. Will the unused 5 million km² of world's tropical forests eventually recover?

Unlike North America in the 1880s, South America, Central Africa, and Southeast Asia offer no large areas of sparsely inhabited, uncultivated, fertile grasslands to lure settlers from the forests. These 5 million km² will probably not be spared the pressure of constant use. Moreover, unlike North America, which was always partly forested, in Cebu, western Kenya, and many other areas the forests are completely gone. There is no source of trees to enable recovery.

Why does deforestation of moist tropical forests result in such bad land that the cleared areas must be abandoned? About a fifth of this degraded land is on mountains. Because of torrential rain, steep slopes, occasional earthquakes or hurricanes, and no trees to bind the soil, it is hardly surprising that landslides carry away soil, fields, homes, and lives. Erosion scars are obvious and frequent on the terraced hillsides of Cebu in the Philippines. On mountain slopes in the Andes or the Philippines, landslides remove not only the nutrients but the soils themselves. Forests may not return, or may take centuries to do so. Forests also afforded protection to the watersheds that supplied water for the richer, flatter agricultural land downstream. The forests' disappearance creates a situation in which everyone loses.

I first saw the fringes of the Amazon basin 20 years ago while traveling eastward across the Andes. The steep hillsides once had extensive rain-forests. The only forests that remained were in the even steeper gullies where a man could not stand up to wield an ax.

The consequences of removing the forests were predictable. We soon came to the first landslide. A bulldozer was scraping a path wide enough for traffic through the rocks and dirt across the hillside where the road had once been. A long line of traffic was stopped on either side. Everyone in the bus piled out and peered downward, inquiring with what seemed to be undue enthusiasm: "Hay muertos?" ("Are there any dead?") The landslides often took buses with them. My guide made it quite clear what he and our fellow passengers thought about those who cut forests so close to the road. Land-slides and the deforestation that caused them were obvious to everyone.

Flying over Haiti, or Madagascar, or New Guinea, where tropical forests meet the sea, blood-red rivers hemorrhage those soils into the blue water, damaging the land and ocean simultaneously. A situation where everybody loses once again.

Even in the lowlands, were cleared forests to be permanently aban-doned, a variety of ecological processes would prevent complete recovery. It is not always true that forests will reclaim large, clear-cut areas. Given the greenery they support, the soils of some tropical forests are amazingly poor in nutrients. The soil around the Amazon clearing of the caboclo I described looked more like white beach sand than the muck we usually think of. Snorkeling in the nearby river one night looking for caiman was a delightful experience. The river was clear, not turbid. Its white sandy bottom reflected our underwater lights in the clear water as would the sand on a Caribbean reef. The nutrients in these forests reside in the vegetation, not in the soil, nor do nutrient-bearing soils erode into the water to muddy it.

In many tropical forests, rich plant growth persists despite poor, thin soils. Many plants have extraordinary adaptations that capture falling leaves. These, not the soil, are a key source of nutrients. Some plants are shaped like large baskets; their roots come into the baskets to adsorb the nutrients from the decaying leaves. Others have long roots that snake across the surface of the forest floor until they find a hoard of nutritious decaying leaves. Burn the forest, and the resulting thin layer of ash is rap-idly consumed by the crops or washed away in the rain. What remain are open, sandy areas that are abandoned within a few years and that may not recover forests for a century.

Cleared areas are often kept that way by grazing them with cattle or goats. At best, grasses do become established, but they form a dense sward into which forest trees cannot penetrate. At worst, goats and cattle rapidly overgraze and trample the vegetation so it rapidly deteriorates, leaving bare areas. Weeds thrive in these conditions. After all, they are warm and wet enough to have once supported forests. The requisite adaptations are obvious: grow well in open, sunny areas with poor soils and be unpalatable to goats and cattle. Across much of the world's tropics, forests have become an almost worthless weed patch. Kenya is in particularly bad shape. Scrawny goats pick among the many inedible weeds for scarce forage. Native trees stand little chance of colonizing, and if they did they would be cut for fuel as soon as they became a meter or two tall.

Tropical Forest Shrinkage

This chapter's first question was, How quickly are tropical forests shrinking? Humid tropical forests have shrunk by 7 million km^2. Most of that deforestation has happened within the last few decades. If all the deforestation had occurred in the last 50 years, the annual rate would be, on average, 140,000 km^2 each year. "On average" is probably not that good an assumption. On the one hand, since some forest losses happened as long ago as the last century the number could be too high. On the other hand, the rate is accelerating so the number could now be too low. Perhaps these cancel out: It does match more recent, detailed estimates.

Myers estimates 121,000 km^2 every year for the 1980s, and this is for the countries that contain almost all of the world's tropical humid forests. The World Resources Institute 1990–1991 *Yearbook* puts the rate of forest loss at 165,000 km^2 each year.[24] These numbers are also quite close to the FAO estimates for 1990 to 1995. For the 30 or so countries where humid forests dominate, their estimate is a loss of about 100,000 km^2. This is the net loss of forest: adding the increase in area of plantations and subtracting the loss of native forest. The total loss of native forest would be higher than the 100,000 km^2. Some tropical countries, such as India, gained forest between 1990 and 1995 because they planted larger areas of trees than they cleared native forest. (India has approximately 150,000 km^2 of plantations, one of the largest areas of any tropical country.)

My best guess from all this is first to take Myers's estimate of the loss of humid tropical forests: 120,000 km^2. Then add to it an estimate of the net loss of other kinds of drier, more seasonal tropical forests. FAO's estimate of these net losses is 26,000 km^2, which again underestimates the loss of native forest. The other sources suggest a loss of drier forests of about 40,000 km^2. These numbers justify the emphasis I've given to humid forests: They suffer the largest impacts. But they are certainly not the only tropical forests to be shrinking from year to year. In summary:

9 Some 160,000 km² of tropical forest are cleared each year.

To make this comparable with the previous numbers, it would have to be converted into so many billions of tons of biomass per year. It is not simple to do so, for reasons which are discussed in Chapter 6. Here, I enter only the area itself.

The next estimate of forest use that I must enter into the spreadsheet is from fire caused by the many people who burn the forest and then move on. Some of this transient use of forests is for agriculture, like the caboclo and the early American settlers. Another part is the deliberate or careless burning of forests as part of other activities.

There is no doubt that the area of fires is much larger than the area of cleared forest. Forests that are left as fragments and edges are also higher than the areas that are cleared. So, too, are the areas that are damaged by logging. These are difficult numbers to estimate, and it is difficult to convert them to an equivalent biomass. I postpone this calculation as well until Chapter 6 and enter this note:

10 Fires burn and selective logging and fragmentation damage areas several times larger than the area of forest cleared each year.

One more number to go. About 7 million km² of tropical forests have been cleared, but in Chapter 3 I estimated the amount of croplands in former tropical forests at about 2 million km². So what happened to the remaining 5 million km²?

They likely appear in global accounts as "permanent pasture," because the dubious economic justification for clearing a lot of tropical forest has been to raise cattle or goats. If so, then this conversion of forests into pastures should appear in our spreadsheet in the same way as the conversion of forests to croplands.

Yet much of this land is just bad land. How much plant production is lost from the conversion of forests to marginal grazing lands and weeds is not an easy calculation to make. It is likely to be a large number because large areas are involved. But calculating by how much less, on average, the former forests produce now versus when they were forests is not so easy.

This is a number that I cannot guess, and so I also postpone discussion of it to Chapter 6. Nonetheless, I must put something into my spreadsheet:

> **11**   At least 5 million km² of tropical forest have been converted to pastures, degraded, or otherwise changed from their original state.

The forest journeys are now almost at an end. I return to Manaus to catch the flight home. For 2 hours we fly northwest and I continue to annoy those wishing to see another movie by looking out of the window.

The view of the forest is breathtaking. I cannot remember being over forest with so little haze. The detail is fantastic. After an hour, we are over the mountains along the Venezuelan-Brazil border. There are occasional tepuis—plateaus with steep sides. They must be high and impressive from the ground for they cast long shadows on the forest beneath. I spot a river that is straight and dropping fast enough to make white water for a long stretch before it starts to meander. What a spectacular, remote place! No wonder that Sir Arthur Conan Doyle, inspired by Victorian explorer Sir Robert Schomburgk's descriptions of tepuis, imagined a "Lost World" with dinosaurs roaming across their flat isolated summits.[25] Then flatland again—the forested lowlands of the River Orinoco and the open areas of the Venezuelan llanos. Some fields are cut into the forest; their neat squares and rectangles betray the first human presence since Manaus. In short order I see Caracas and beyond, the Caribbean.

As we cross the water, I marvel at the last couple of hours. To travel that long without obvious signs of human interventions on the landscape is a special experience. The answer to this chapter's first question is that this experience is becoming ever more special as forests shrink at the rate of 1.6 million km² each decade.

As I was leaving the forested banks of the Río Negro, the caboclo had struck the match to ignite the trees he had earlier girdled and killed. He and all the other modern-day Loges burn an area of tropical forest much larger than the area of forest cleared each year. He and his companions in Peru, Burundi, or New Guinea will not be repeating our histories. The farmland he creates will not look like Western Europe with its neat productive fields, trimmed hedgerows, and small coppiced woods. Or like Japan, where bright green rice paddies abut new factories and high-rise apartments. It will look even less like the eastern United States, where forests and fields intermingle, or that continent's Midwest where grain fields stretch unbroken to the horizon. We are long-time settlers in what was once forest, but we have saved the best forests for our use. Our history does not predict the Amazon's future. The smoke that now rises from the caboclo's forest is a harbinger of rapid destruction and transient occupation. Recovery? Perhaps, but only if the field I had walked across is left alone to be reclaimed by the surrounding forest.

Wagner's opera, by the way, does have a happy ending, if you can wait the 15 hours or so that it takes for the music to end. After Brünnhilde and everyone else are dead, you hear the Rhine reclaim its magic gold from Valhalla's ashes and the cycle begins anew. The growing throngs of tropical poor will not be able to leave the land alone after they clear the forests. The cycle will not repeat. They have nowhere else to go and must make do with less than the best. What happens when we cannot move on?

5

Peace and Quiet
and Good Earth

It was March 1774, and Captain James Cook needed to land his ship. He was recovering from an illness that had nearly killed him, and he and his crew were desperate for supplies. One Easter day 50 years earlier, the Dutch explorer Roggeveen had found an island nearby, so Cook headed for it. On landfall, what Cook's crew found so alarmed them that he was "determined to quit the island without further delay," even though the next landfall was three weeks away. He had encountered a land that people did not have the means to leave.[1]

At a latitude that should have afforded pleasant, fertile land, the island was treeless, houses and crops were few, and the ground was "dry, hard clay." It was "a most desert, barren country full of rocks." The landing party was able to trade for a few sweet potatoes, but only until the fields' rightful owners came along, since the traders were stealing produce they didn't own. There were a few chickens and a few terns, whose eggs they would have eaten gladly, a few rats—the kiore, a Polynesian delicacy—but the locals would not trade them. Worst of all, there was almost no freshwater. What was available to the sailors was "stinking" and too salty or too filthy to drink.

The sailors were not greeted by hundreds of people paddling their outrigger canoes, so familiar a sight on other Polynesian islands. There were only "three or four canoes on the whole island and these very mean, made of many pieces of boards sewed together with small line." So barren was the landscape, Cook surmised that they had scavenged driftwood to build them. The islanders lived in "low miserable huts." The officers who went ashore guessed that from 300 to 500 people lived on the island. They saw fewer than 10 women.

Yet this miserable, remote scrap of land, almost 4000 km from South America, hosted more than 200 stone statues. Seven hundred more were in various stages of being quarried. The statues stood 10 m high and must have weighed nearly 100 tons each.

Who made them? Such monuments would require labor and a social structure that could organize such a project. The islanders paid scant respect to the statues, and many were toppled, broken, and defaced. And how were the statues made? Where were the materials for ropes to pull them upright and to roll them across logs to their platforms? Some stood 10 km from their quarry sites. Where were the logs?

The mysteries of Easter Island's story had to wait for archaeological exploration. From plant pollen, we now know that prior to settlement a lush subtropical forest covered the island. The most common tree was a palm, closely related to one on the South American mainland. That palm grows to 25 m, and has edible nuts and a sugar-producing sap. Other common plants on the island would have provided firewood and fibers for ropes.[2]

From discarded animal bones, we know that there had been a huge seabird colony on Easter. It was one of the largest in the Pacific, with albatrosses, boobies, frigate birds, and terns. The colonists ate the seabirds to extinction, sparing only the terns that nested on hard-to-reach islets offshore. They completely deforested the island, driving the palm tree to extinction. Only a few small specimens of the bushes that produce rope and firewood survived. Without this protective vegetation, soil quickly eroded. The once lush land was stripped of soil to bare rock. Streams dried up.

When the palm went, so did the means to erect statues and materials to build canoes that could have been a means of escape. Worse, before they lost their canoes most of their meat was from the sea—some fish, but more porpoises. Oral tradition and fragments of bones in caves suggest that without canoes, the search for meat meant capturing men. The victors imprisoned them, fattened them up, and had them for dinner. Cook's party saw few women because they were hiding in the island's many lava caves for protection.

The Easter Islanders, who once had free time to build their extraordinary statues, starved—perhaps 90 percent of them. On another remote South Pacific Island, Pitcairn, the story was worse. All the island's people died.[3]

Living on the Land

Land does not guarantee the means for existence. Living in the southern United States, as I have done for half my life, one quickly learns to answer the question "Who are your people?" with its implicit command to defend the gentility and, hence, antiquity of one's ancestors. "Those on my father's side came over by boat," I reply with the answer of so many Americans whose families arrived in the great exodus of Europeans a century ago. In "The South" that's not long enough, but I tease them. My people crossed the North Sea to England, not the North Atlantic to New York Harbor. By the time the Domesday book counted my ancestors in 1085, we had been on the land for 500 years. The Celts on my mother's side consider the 50 generations of Anglo-Saxon farming folk to be newcomers.

Jane Austen's novels or the statistic that only 15 percent of English place names have disappeared in the 900 years since the Domesday book cataloged them, all conjure up the same dreary image of agricultural persistence.[4] Farming the same fields for millennia, Britons are like the hobbits of Tolkein's Middle Earth, a "very ancient people" living on "a good-tilled earth." For many of the world's farmers, that seems as improbable as hobbits.

"Land, Katie-Scarlett! It's the only thing that matters! It's the only thing that lasts!" cries Scarlett O'Hara's father in *Gone with the Wind*. Yet it wasn't even true of the antebellum South of the 1860s when he said it. The O'Hara's were rich and had the best bottomlands with deep soils. Poorer families were already abandoning their depleted hillsides and moving west. Like Steinbeck's poor family, the Joads, in *The Grapes of Wrath*, they would deplete the soils in eastern Oklahoma too, and move yet again. They would leave few traces other than the ecological damage they caused.

The State of the Land

The answer to whether your people are O'Haras, Joads, or hobbits depends on how well you have managed your land's biological investment account. This chapter's first questions are, What fraction of the world's land is degraded and by how much? By how much have we reduced the principal so the biological interest it generates is reduced? And to what extent has humanity wrecked the land to the point where it must be abandoned and thus generates no interest at all?

Are the grisly stories of Easter Island and Pitcairn warnings to us? For we can no more escape our island Earth by spaceship than the Polynesians could their islands. To what extent have we reduced the land's ability to support our activities so that we must move on, like the Joads? Which land will support ancient people, which will we totally destroy, and which will we merely diminish its fertility? And where are these different classes of land?

The ruined Mayan temples of Tikal, Uxmal, and Chichen Itzá rise from the forests of Central America. The entire landscape was once dotted with settlements; those around the temples were huge. The same is true for Angkor Wat, in modern-day Cambodia, built by the Khmer. There are the abandoned Anasazi settlements of the desert in the southwestern United States. In the first few centuries BC, Ptolemy and Socrates knew that the land west of Alexandria, in Egypt, was rich and populated; today it is mostly barren.[5] Examples of abandoned land span Earth's varied ecosystems.

In Chapter 4 I noted that 5 million km^2 of degraded land are in the former humid tropical forests of the world. In view of this large area of degraded forest, the answer to this chapter's first set of questions is remarkable: Humanity has diminished plant production over a large extent of the drier, nonforested land surface. There is less plant growth to use, and it is easier to abuse what there is.

This chapter has a dual purpose: to estimate land degradation and to cover the history of the drier parts of the planet. This dual purpose begs a second set of questions. Why are drylands so vulnerable? And why do we manage them so badly?

Although a square kilometer of desert, grassland, or dry woodland does not produce as much plant growth each year as does the same area of forest, their combined area is huge and they make a major contribution to the planet's total plant production. Drylands constitute roughly half the Earth's land area. As in the previous chapters, we encounter problems of definition: Quite what the *dry* in *drylands* means is arbitrary. According to a UNEP report, drylands constitute 61 million km^2.[6] Jerry Olson's total is also 61 million km^2, which, given the problems of definition, is closer than I have any reason to expect.[7] That's half of the ice-free land surface.

Like forests, *drylands* are quite diverse. At the wetter end of drylands are dry woodlands and savannas. In dry forests, there are trees but the canopy is not closed; space lies between the trees. This distinction separates them from closed canopy forests. In savannas, the trees are widely scattered. In Olson's data, these forested drylands account for about 15 million km^2 of his total (see Map 8).

Grasslands are the most widespread class of drylands. They make up about 22 million km^2 and are dominated by grasses. Some are converted from other kinds of areas, including forest. Grasslands grade imperceptibly into drier, treeless areas, but ones where shrubs begin to dominate. Areas dominated by shrubs include moist Mediterranean ecosystems found around the Mediterranean Sea, but also in California, Chile, South Africa, and Australia. These drier but still shrub-dominated systems constitute another 3 million km². Map 9 combines both grasslands and shrublands. In turn, these areas grade into even drier areas that include more than 11 million km² of semidesert and more than 9 million km² of desert (see Map 10).

Shrublands, semideserts, and *deserts* are all dominated by shrubs rather than grasses; the vegetation becomes sparser and the grasses less important as the rainfall decreases. Where one draws the line or whether one invents other gradations—"desert grasslands" or "shrub steppes"—is a matter of personal preference. On starting graduate work in southern New Mexico, I asked where the deserts were: The area looked too lush after my undergraduate experiences in Afghanistan and Iran. Yet I have met people for whom driving on New Mexico's Interstate 10 was the adventure of a lifetime. I did not tell them that most of this journey is across a lovely grassland, albeit a rather dry one. It is not a desert, for grasses outnumber shrubs. And, no, New Mexico does not look like the Sahara. Although that journey may seem an adventure after the dense settlements east of the 100th meridian, there are towns every 100 km and ranches along the entire route.

We at least try to live in drylands. What are the consequences of our doing so? While most of the world's crops grow in areas that were once forests, drylands grow crops with and without irrigation. They also support grazing livestock and animals hunted for game. Relative to their area, drylands contain a smaller fraction of the world's total croplands, a larger fraction of those that are irrigated, and a much larger fraction of grazing lands. The biology is quite simple. Crops grow better when there is adequate water. If water is insufficient, it has to be supplied. If drylands cannot grow crops, then they may support cows, sheep, goats, or camels. Dry woodlands and shrublands provide fuel for fires.

Desertification

All these human activities can—and often do—lead to the degradation of the land that is one of this chapter's two themes. For simplicity, I lump them into an all-purpose category called "desertification." The UNEP report mentions the most obvious symptoms:

> *Reduction of yield or crop failure in irrigated or rain-fed farmland; reduction of perennial plant cover and biomass produced by rangeland and consequent depletion of food available to livestock; reduction of available woody biomass and consequent extension of distance to*

sources of fuel wood or building material; reduction of available water
due to a decrease of river flow or groundwater resources; encroachment
of sand that may overwhelm productive land, settlements or infrastruc-
tures; and increased flooding, sedimentation of water bodies, water and
air pollution.[8]

Like other UN documents, you should not try to read this sentence
aloud without taking a deep breath. But you get the idea. We do a lot of bad
things to drylands.

A former colleague of mine, Harold Dregne at Texas Tech University's
International Center for Arid and Semi-Arid Lands Studies, generated the
report's numbers. Texas Tech is in Lubbock, a town on the high western
plains, surrounded by irrigated cotton fields, and close to cattle ranches. It is
as good a place as any to understand drylands.

Dregne's approach was similar to the one that Myers employed for
tropical forests. He assembled more than 600 articles from around the
world and talked to people with local experience. Then he came up with a
country-by-country assessment of the extent and severity of dryland degradation.[9]

To make some sense out of drylands and the processes that degrade
them, Dregne split the problem into three parts: the fate of irrigated crops,
rain-fed crops, and grazing lands. To his list, I add the dry woodlands from
which many Asians and Africans get their fuel. These parts suffer different
impacts, and each continent differs in its susceptibility to them.

Nearly 2.5 million km^2 of croplands are irrigated—or about 15 percent
of the planet's total. Over 60 percent of this irrigated land is in drylands—
about 1.5 million km^2. The rest includes areas such as wet rice paddies. Irrigated drylands annually receive less than about a third of a meter of rain
each year—much too little to grow crops without additional water. Add
water to a hot dry climate and plants grow fast. Irrigated land is disproportionately important because it produces about 30 percent of the global food
harvest. That is twice as much as one would expect for its area.

These lands are always a surprise. Cross the desert on a long, dusty trip,
hazy in the heat, everything colored buff. Then we see irrigation, a pump
showering water high into the air making it suddenly humid and creating a
border so sharp that you can stand with one foot in desert and the other
among dense, dark green crops. From the air we see a river—the Nile or the
Río Grande—a long, thin snake with square markings that delineate its
fields sharply from the adjacent barren desert. Or a metal rig revolving

around a central pump, spraying water into the air and forming neat circles of crops.

Of this irrigated land, 30 percent—an area roughly the size of California—is moderately to severely degraded and that percentage is increasing. Certainly, some dryland areas have been irrigated for millennia. Those in the Nile Valley are more than 3000 years old. Along China's Yellow River as it loops northward from Yinchuan toward Inner Mongolia are irrigated lands more than 2000 years old. Other areas are not so lucky.

Although the numbers are hard to substantiate in detail, the UN Conference of Desertification suggested that we lose about 1 percent of the irrigated cropland each year from the world's drylands. (This percentage is also the one suggested by David Pimentel and his colleagues for the rate at which arable land is abandoned worldwide.[10]) If 1 percent does not seem that much, stretch that out over an average human lifetime; it is over a lifetime that people have to eat!

On irrigated drylands, the main cause of this destruction is salt accumulation. There is one huge practical difference between rainwater and the water that runs off the land to be used later for irrigation. *Rainwater* is the condensation of seawater distilled by the sun. Runoff has percolated through the soil and has dissolved salts there. Lots of chemical compounds are technically salts, but *the* salt, sodium chloride, is by far the most common. That is why the sea is salty, for it is where the salts end up.

When the runoff evaporates it does not take the salts with it. On irrigated land an average of 80 cm of salt-laden water evaporates year after year.[11] So salts accumulate every year, unless yet more water is provided to flush them out. When these salts are left behind, productive land can quickly become barren salt flats, with nothing but a few scattered halophytes (plants adapted to high levels of salt in the soil). Look for irrigation in hot areas and that is where you will find massive losses of land owing to salt accumulation.

Asia and southwestern North America account for 75 and 15 percent, respectively, of the irrigated land that is degraded by salt. The worst affected country is Iraq: Over 70 percent of its irrigated land is degraded. In Russia, where the Volga River runs into the Caspian Sea, the irrigated lands may last only a few decades before the accumulating salt destroys them. In the United States, salt accumulation has lowered crop yields across at least a quarter of the nation's irrigated land, or 50,000 km². Mexico has diminished yields across a third of its irrigated land. These numbers provide this chapter's first entry into the global spreadsheet.

> **12** Salt accumulation destroys about 15,000 km²
> of irrigated land each year, and it diminishes
> the potential biomass production across 450,000 km².

Drylands also support rain-fed crops. Another 5 million km² of cropland receive less than 60 cm of rain each year. These lands, nearly a third of all croplands worldwide, are at the margin of what we can exploit.

At the margin, they are especially vulnerable, and we damage them accordingly. More than 2 million km² of rain-fed croplands are moderately to severely degraded—almost half the total. Each year about 1 percent of these lands is abandoned. Those numbers are the next to be entered into the spreadsheet.

> **13** Agricultural practices destroy about 20,000 km²
> of rain-fed cropland each year, and they diminish
> the potential biomass production across 2 million km².

This proportional loss of rain-fed croplands is similar to that of the irrigated lands, but the causes are different. Wheat, maize, and the other grains that constitute 85 percent of food production are annual plants. Annuals grow from seed and die within the year. Harvesting annuals leaves the soil bare and vulnerable to erosion by wind and rain, thus destroying the soils and depleting them of their nutrients. By contrast, native grasslands are mixes of mainly perennials. These grasses live from year to year and have deeper roots to bind the soil. They are better buffered against the vagaries of nature.

Dust Bowls

Just what wind does to bare ground in dry regions of the world has to be tasted to be believed. Whenever I think of my years in Lubbock, I remember the dust storms. The fine grit seeped into everything, including food, to which it added its distinctive flavor.

The song that goes "Home, home on the range, where the deer and the antelope play" was written near Lubbock. Not anymore. The deer and antelopes were replaced long ago by cotton fields for as far as one can drive along the straight west Texas roads. It was Winter 1977, and at least the song's last line was right: "and the skies are not cloudy all day." It was a brilliant winter's day with low humidity and bright sunshine. The jet about to land at the airport came in over my house. Like Mary Poppins, it alerted me to some imminent change in the weather. This was not its normal flight path. Looking to the west, just above the neighbors' rooftops, I could see a low cloud bank, but it was a dull orange, not white. The cloud wall grew steadily, until it soon towered thousands of meters above my head. "The orange blob," I thought, inventing the title to match the opening scene from a cheap science fiction movie. Then the dust hit: no rain, just fine orange silt that, like a thick fog, obscured all but the houses nearest to mine.

I traveled east that December for the holidays, first in the storm and then on its heels, across Oklahoma and Arkansas. In Oklahoma, it was the worst dust storm in 20 years.[12] Long before I saw the satellite imagery that confirmed it, I knew that the storm had started on the north-south border between Texas and New Mexico. On the eastern, Texas side, from autumn to late spring, wind whips up the soil of the bare cotton fields. Yet if you cross into the New Mexico grasslands, you quite suddenly encounter antelopes, deer, and clear, dust-free skies.

Drylands do get rain and, when it comes, it can be violent. Storms can drop several centimeters of rain in an hour. They can also drop golf-ball-size hail that can smash crops, pulverize bare soil, and, as I experienced more than once, damage the roof of my Lubbock house. Unprotected by perennial plants above and below ground, rain-fed croplands in dry areas are eroded by wind when they are dry and eroded by water when they are wet.

These examples are misleading in one respect. Harold Dregne estimates that only 16 percent of North America's rain-fed croplands are damaged. In Africa and Asia, wind and water erosion has damaged over 50 percent of these lands. Winds blow African soils into Florida and Brazil each year, and soils from China end up in Hawai'i each spring.

Madagascar gets its nickname—the red island—from its water erosion. Its rivers are blood red from the sediment that washes down from the land. I had seen that hemorrhaging on satellite images long before I boarded the plane to Antananarivo, the country's capital. Flying along its western coast, I could see close up what the satellite images showed from afar—the red streaks braiding with blue waters in the estuaries. Over land, the cause was obvious—hardly a scrap of forest remained—and on landing: a visual shock. The landscape looks bone dry, the yellowed grasslands with few trees resembling a western U.S. state, like Oklahoma, I thought. Yet amid this aridity are paddy fields, lots of them, wet places with abundant domesticated ducks paddling over them. Much of Madagascar is wet, the forests absent because they have been cleared again and again by fire.

Rain and insufficient vegetation to bind the soil, and the resulting erosion is inevitable and expected. There is one more visual shock, the lavacas. These are huge erosion pits, a couple of hundred meters deep in places, where the surface has broken and the unimpeded rain erodes the soils and rock beneath into steep-sided cliffs and gullies. Driving across the Malagasy landscape, drowsy from yet another long day in the Land Rover, I imagine the land as some giant piece of yellow cheese, with the scattered lavacas as the bites from the incisors of a giant mouse.

Grazing the Drylands

In Chapter 3, I entered the number of 2 million km^2 of grazing lands that are converted temperate forests.[13] These lands are highly productive ones, with bright green fields and lots of animals on them. Combined, these grazing lands produce about half the global production of livestock.

The other half of the livestock are on the much larger area of grazing lands that were not converted from something else. They are the natural grazing lands where our domestic animals share the fodder with their wild counterparts. While these lands include that grazed on by Arctic reindeer, they are mostly made up of drylands. There is considerable disagreement as to how much land should be counted as wild grazing land. One person's grazing land is another's barren desert.

There is a huge range of livestock densities. New Zealand, famous for its sheep, has 380 of them, accompanied by 60 cows, on an average square kilometer of its pastures. The veldt of Swaziland and adjacent South Africa supports about 60 cows and 40 sheep per square kilometer. Afghanistan has few cattle, and its sheep and goats number only about 50 per square kilometer. In the Arabian peninsula, the density of sheep and goats drops to less than 10 per square kilometer. So, New Zealand and Saudi Arabia illustrate a continuum of grazing lands that differ by a factor of 40 as to how many animals they support in an area.[14] Dregne counts 46 million km^2 of dryland grazing—a number that is larger than some other estimates. Dregne's point is that only the most extreme deserts and the coldest tundras are not grazed at least occasionally. Most areas in the world will be grazed when conditions are right, or when the graziers are optimistic enough. Let's visit some grazing lands.

According to the FAO, there are nearly 16 million sheep and goats in Afghanistan. That number does not surprise me, though how anyone counts or estimates sheep in such difficult and remote terrain mystifies me. FAO also reports that 300,000 km^2 are permanent pasture—a figure simply rounded to the nearest 100,000, perhaps an honest statement of uncertainty. That area is almost half the country. As such, it has one of the highest proportions of pasture of any country in the world. Only Kenya, Somalia, South Africa, Mongolia, and Uruguay exceed it.

For Afghanistan, if we assume an average plant growth of 200 tons of biomass per square kilometer per year,[15] the country's total forage production is about 60 million tons. Using Morrison's numbers (see Chapter 2), we find that the Afghan sheep and smaller numbers of camels and other livestock consume only about 3 million tons, or 5 percent of what is available.

Why do livestock take so little of the annual plant production? Between Herat and Kabul, the paved road east to west across Afghanistan takes a long loop southward through Qandahar. Permanent pasture? Each evening as we stopped for the night, the fantasy of soft patches of grass eluded us. We did our best to sleep on the bare, stony ground. Before dawn, in the coolest air of the day, we would hear the shepherds and the bells of their fat-tailed sheep that were already out looking for the scant vegetation between the rocks. Grass eluded them, too.

Watching sheep and shepherds look for the last green foliage on a July morning explains why the sheep consume so little of the land's annual plant growth. The temperature that day rose to nearly 40°C. No rain had fallen since late March, nor would any more fall until December.[16] No wonder that by midsummer the landscape looked so barren. Only after the winter rains would the land show a thin wash of green. Much of the land's growth comes in too short an interval for the sheep to use, though they store fat in their tails to ward off later starvation.

Another road: This one is along the border of Swaziland and South Africa, at another time—October, which is early spring there—and another use of the plant production. We passed a hillside colored such a soft, bright green that it could easily have been the English Pennines. The grazing sheep heightened that impression, but they were Merinos, not bare-faced Cheviots. The impression was dashed at the field's edge; for stretching 2 km or more over the hill, a track separated the field from its blackened neighbor. In that field the ground was so completely burned that only black soil, strewn with granite boulders, remained. Yet, in the charred tufts of the grasses were numerous tiny, green shoots. A third hillside was yellow with dried grasses. Green, black, and yellow alternated across hundreds of kilometers as we drove eastward. From the air, the colors following the capricious path of these fires looked like the splotches of a small child's painting.

Three weeks later, the burns were splashes of bright green new grass. Among the grasses were showy flowers. The grass roots survive the fire, and, as rain dissolves the nutrients in the fires' ashes into the soil, they can quickly absorb them for new growth. Fire is a natural process in the African veldt. After a long dry season the first electric storms often ignite fires; fire and rain go together, and the grasses are adapted to them. But these fires are not natural ones. Graziers burn them in controlled areas to improve grazing. In calculating human impact on natural grazing lands, we have to count the amount of plant production that goes up in smoke and into ash from fires that are deliberately set.

How much of the world's permanent pastures are burned this way every year? The Stanford group quotes a figure of 6 million km^2—a lot of area—but only a sixth of the more conservative estimate for the total area of permanent pasture. I have no way to check this figure, though it seems plausible from my experiences in Africa and elsewhere. I enter the area in my spreadsheet.

> **14**   About 6 million km^2 of pasture lands
> are burned each year.

A final question: How much of the plant production of these grazing lands have we destroyed? Worldwide, the damage to rangeland is of two kinds: damage to the soil and damage to the vegetation. Dregne estimates that about 10 million km^2 have soil damage, 27 million km^2 have vegetation damage, and about 8 million km^2 have both. About three-quarters of the 46 million km^2 of grazing lands have diminished plant growth.

> **15**   Grazing has degraded biomass production
> across 35 million km^2 of dryland.

What does Dregne mean when he concludes that so much of the world's grazing land has "damaged vegetation"? Grazing livestock eat the vegetation, sometimes down to the ground. Livestock walking to and from permanent water sources trample plants, destroying them and their roots that bind the soil. When the rains do come, it is along such tears in the fabric of the prairie, steppe, or veldt that rivulets may form and wash away the soil. The largest total damaged areas are in Asia and Africa, but the largest proportion (85 percent) is in North America. Let's travel westward across that continent.

The East's deciduous forests are thick and green until we reach the Mississippi River. We cross the river at Memphis, Tennessee, turn slightly south of due west to take the southern route through Little Rock. The land is drier in Arkansas, but there is still too much forest to believe that we're in "The West." Cross the Red River and we're there: Texas. The very name summons the cattle, and the hats, horses, and pickup trucks we need to manage them (or at least pretend we do).

For 1000 km the road heads westward through Dallas, Abilene, Midland, and ever drier country. By El Paso and the New Mexico border, desert shrubs dominate the dry hillsides. The road climbs out of the lush, irrigated valley of the Río Grande—a series of muddy puddles for much of the year—and heads west to Arizona. Mountains intersperse the mix of grasslands and shrubs for another 500 km. Then the road drops down into Tucson and its characteristic saguaro cacti. These are the driest and most vulnerable of North America's grazing lands.

The evidence of their destruction can be as obvious as a fence line with grass on one side and dense shrubs on the other. In the clear, dry air of America's Southwest, that fence can stretch to a distant horizon. Grasses make grasslands, but woody shrubs, cacti, and an assortment of other inedible or poisonous weeds destroy them. When livestock overgraze the native vegetation, they open a niche for invading shrubs that livestock cannot eat.

In Texas alone, more than 300,000 km^2 of range land—roughly half of the area of this large state—are infested with mesquite and other invasive woody shrubs.[17] The mesquite dunes, clumps of the low shrub separated by open areas of red sand, stretch for day-long car journeys, across Texas, south into Mexico, and east into New Mexico. These lands, classified as permanent pastures, have stocking rates of just a few animals per section (a square mile, roughly 2.5 km^2). That stocking rate appears to be too low to make any practical economic sense and, as evidenced by the invasive shrubs, too high to manage the range sustainably.

In other parts of the world, the processes are similar; only the names of the offending weeds are changed. Lantana, a short, unpalatable bush, is a popular garden plant. From gardens, it jumped the fence and now dominates large areas of former grazing land in areas as diverse as Africa, Australia, Hawai'i, and South America. Gardeners spread opuntia cactus, with its spine-defended fleshy pads, in the same way. It once dominated huge areas of northern Australia until brought under control by an insect that would eat it. It is still a problem in many areas.

How and when was the American West lost? For Arizona we know the answer—invading mesquite in the 1880s—and have photographs to prove it. The source is *The Changing Mile*, a remarkable book by James Hastings and Raymond Turner that compiles new and century-old photographs.[18]

Before the photographs, Hastings and Turner recap the history. Within a decade of Cortés's conquest of the Aztec Empire in 1520, Spaniards were on the Pacific Coast at the southern end of the Gulf of California. Mission by mission they moved northward, each donating herds of cattle to the next. The role that these cattle and ten times more numerous sheep played in the ecosystems is not entirely clear. Ranches certainly suffered more than the occasional setback at the hands of raiding Apaches. By 1854, when Santa Ana yielded the land north of the present United States–Mexico border, ruined haciendas and wild cattle were common sights in what was to become Arizona. Hastings and Turner concluded that cattle were common, but not enough to change the landscape. Contemporary accounts are of streams running through open land with plentiful grass.

Without a war to preoccupy them, the U.S. Army could close in on the Apaches and did so in the 1870s with great success. Settlers came in under their protection and the railroad arrived in 1880. Arizona Territory held 5000 head of cattle in 1870 and 1.5 million by 1891. A cow was worth $5 at birth. Grazed on public grass for five years, its value escalated tenfold. Grass was gold, and cattlemen grew rich.

In 1891 and 1892 the summer rains failed, and by early 1893 dead cattle lay everywhere. Perhaps three-quarters of the herds died. Thousands of square kilometers of grasslands lay stripped of their plant cover, exposed to the elements. Photographers, accompanying the surveyors of the international boundary, lugged along their heavy plate cameras and recorded the scene.

It is these photographs that Hastings and Turner assembled and then returned to the exact locations to compare them. The "mile" of their title refers to the range of elevations of the region, one large enough to encompass many kinds of vegetation: desert, grassland, and oak woodlands. The desert photographs are often strikingly similar, some with the same individual saguaros visible, now 70 or more years older. The grasslands have changed the most and the most consistently. Thickets of mesquite replace open grasslands in photo after photo.

There was another change, too. Along the U.S. border, the 1891 surveyors, their horses, and their dog look across the threadbare prairie at a stream. Apart from the sparser grass, it looks like a description of the place that had been written 34 years earlier. "The descent is made by a succession of terraces. Though affording no great quantity of water, the river is backed up into a series of large pools by beaver dams and full of fish." In the 1960 photograph the grasses are gone and the watercourse is deeply scoured with steep sides. "Arroyo cutting," as it is called locally, and the flooding that causes it, started in the drought of the 1880s. The loss of vegetation opened up the soil to erosion. Water cut deep channels into what was once gently sloping prairies. As if this were not bad enough, the deeper channels allowed more water to run off the land, so less water remained in the soil to support already stressed grasses.

Retirees who flee the cold winters of their northern home states to spend their last years in Arizona may not care anymore about mesquite invasion and arroyo cutting than they do about their taking the Colorado River's water for their well-watered lawns and golf courses. In Africa, however, these processes of declining plant production have far grimmer consequences.

Drylands in Africa comprise 20 million km^2, some 65 percent of the continent.[19] One-third of Africa's drylands are arid deserts, uninhabited except for a few oases. The remaining two-thirds of the drylands also support two-thirds of the continent's population, one that is rapidly rising. As the population grows, the land degenerates. Dregne estimates that 18 percent of the irrigated cropland, 61 percent of the rain-fed cropland, and about 75 percent of the rangelands are at least moderately harmed by desertification.

Recurrent droughts are a permanent fact of life in Africa's drylands. Almost every year there is a drought somewhere, and the major droughts regularly affect larger portions of the drylands. As in Arizona a century ago, a rise in population and an increase of livestock can run headlong into a drought that has catastrophic consequences. Such events occurred in the years 1968–1973, 1982–1985, and 1990–1991. During these years many African countries experienced substantial food shortages. As in Arizona, unsustainable grazing and droughts damage the land; grazing animals must move to even less suitable land; and desertification can worsen with each drought cycle.

From space, the satellites that survey the planet every month show the seasonal ebb and flow of plant growth. In North America, Europe, and Asia the land turns green as it warms and spring moves northward, and then turns brown again in the autumn. In Africa, north and south of the always green rainforest, the green follows the seasonal rains. The area of this green pulse varies greatly from year to year; its rainfall-dependent northern and southern boundaries fluctuate. Is there an overall trend? Is the Sahara expanding? The popular press certainly thought that it was.

Jim Tucker and his colleagues have examined this question too. They used satellite data and ground rainfall measurements to study variations in the size of the Sahara Desert from 1980 to 1997. They mapped the "shoreline" of the desert "sea"—the boundary that receives 20 cm of rain each year that has desert on the drier side and the Sahel on the wetter. It moved considerably from year to year. The area of the Sahara Desert varied from 9,980,000 km^2 in 1984 to 8,600,000 km^2 in 1994. While they concluded that there has been almost no systematic change in the rainfall, nor in the position of the southern desert's "shoreline" from 1980 to 1997, the fluctuation of almost 1.5 million km^2 in the size of the desert is devastating to those who live in its path.[20]

Consuming the Land

There is a paradox in these figures about the loss and degradation of cropland and grazing lands. They seem to contradict the year-by-year global agricultural summaries from the FAO. The losses run at roughly 1 percent in each of the three categories: irrigated crops, rain-fed crops, and grazing land. Yet look at the FAO numbers. Worldwide, the land under crops increased by over 1 percent in the decade prior to 1991, the percentage that was irrigated grew by 2 percent, and permanent pastures rose by nearly 4 percent. How can this make sense?

The answer is that we move across the landscape, consuming it. Irrigated lands are lost as salt accumulates; we compensate by building new irrigation schemes to flood what were once rain-fed croplands. This area then decreases as we abuse the soils. To compensate for these losses, we convert the best grazing lands into rain-fed agriculture, expanding grazing onto land previously considered too poor. This is not sustainable land use. We continuously destroy the best land and move to poorer, drier lands.

The more arid the land, the higher is its rate of desertification. A case study of Tunisia, for instance, found that in the century before 1980, cultivated land increased from 6000 km^2 to 36,000 km^2, a net gain of 30,000 km^2 mostly at the expense of grazing lands.[21] Grazing lands, however, shrank by 40,000 km^2 because 10,000 km^2 were also lost to the Sahara. On desert fringes the rates of desertification are particularly high. Unlike the world average, here the grazing lands are shrinking; there are no more "poorer, drier lands" to consume.

Averaged over many such case studies, the UNEP reports show a hundredfold difference in the annual rate of desertification in different parts of the world. The annual rate of desertification is roughly 10 percent per year in arid lands (such as Tunisia), 1 percent in semiarid lands, and a tenth of 1 percent in the wettest of the drylands.

Despite the uncertainties in the global statistics the general pattern is clear. The 4 percent rise in the total area of grazing lands over the last decade is not a measure of progress. It is more a sign of incipient disaster, of our need to put livestock on ever more unsuitable land.

Why are we so inept at managing the land on which our very survival depends? It is not just drylands—for we damage forests, too—or even land. As we overgraze the land, we overfish the oceans. *Overgrazing* and *overfishing* mean that we have harvested something, grass or fish, below a maximum level that might sustain us. We kill so many fish that there are not enough to reproduce. Why don't we adjust our use of grass or fish according to what they produce?

Our lack of prudence can be explained by our behavior, our societies, and our economic choices. I postpone those explanations until after Chapters 8 and 9, which tell of the oceans and the case histories they provide. On land, it seems to be especially drylands that we harm. The ecology of drylands provides an ecological explanation. It involves the word *adjust* in the preceding paragraph. Nature is variable. Our response to it isn't. Often, we do not or cannot adjust.

Rainfall's Bad Bet

On grazing lands it is rain that drives that variability. Plant growth varies in concert with the rain; the number of animals that the land can support must vary as well. Stocking too few animals in a wet year means throwing money away. Stock too many in a dry one and they will starve. And as they starve, they harm the plants and reduce plant productivity for years to come.

How variable are rains? Home from the university one December in the early 1970s I sat watching the news with my parents. The announcer had read the litany of depressing news in the pompous tones that the BBC was then using to suppress the variety of English accents. Following the standard model, he was about to end the broadcast with something amusing. "Today, it rained in the Qattara Depression in the Libyan desert." He paused for effect and looked at his notes. "So what?" Dad retorted in his very different accent. "It rained when I was there in 1942." (He had been at the desert's eastern end near a small, obscure rail stop called El Alamein.) "The last time it rained in this area . . ." the announcer continued, "was in 1942."

Deserts are like that. For years, decades sometimes, they remain barren and seemingly lifeless. Then, against all odds, clouds make it across the sheltering mountains or the long distances from the ocean that usually strip them of their water. It rains. Within days, the desert flowers, the birds sing and nest, and the frogs emerge from underground burrows to spawn in temporary ponds. Through their evolution, all understand the urgency of propagating the next generation. It may be a long time before the next chance.

Grasslands are like this too, though they are neither as dry nor as uncertain as deserts. There is gradual change. As areas get drier, the rain becomes less certain from year to year. Scientists measure this uncertainty by an index of how variable the rain is from year to year relative to the amount of rain in an average year.[22]

Earlier we traveled eastward across the United States. Suppose that we had picked up the long-term rainfall data from weather stations as we went. Data for the summer months are what matter, for that is when the plants grow. East of the Mississippi the monthly rainfall would have a variability index of about 50 percent. Such a number would be generated by a series of years with an amount of summer rain in half of those years and twice the amount of summer rains in the other half. Some years are wet. Some are wetter.

By the time we reached the most eastern prairies of Texas, the variability index would have risen to 100 percent. Such a number corresponds to a wet year's having about five times as much rain as a dry year—almost exactly the ratio found in Lubbock, Dregne told me. By Arizona and the land of *The Changing Mile,* the variability would be as high as 200 percent, which is equivalent to rain falling in the summer of one year and no rain at all in the next three summers. Travel on to Las Vegas, Nevada. The variability would be 400 percent. It rains one summer in 16 years. Slot machines are a better bet than summer rains.

Wet areas are consistently wet. Dry deserts are so infrequently wet that no one is fooled into thinking they can be settled. In between, grasslands and other drylands offer a deadly illusion. In a good year a grassland can look magnificent: horizon to horizon of lush green grass, its larks singing, and its antelope fat. Settlers on dryland: Beware of the good year and the good life it promises. There will be the odd good year, but there will be more bad years.

And now for a result that my seminar audiences often greet with incredulity.[23] Rainfall is more malevolent than gambling with dice. Settlers: Beware the good *years.* Note the plural! If only one year in six is a good one, surely two good years back to back would be as likely as throwing two sixes in a game of dice, a 1 in 36 chance. Not so. Good years come together more often than we expect. Once you finally get a six, you have a better than 1-in-6 chance of getting another six on the next throw. Unfortunately, bad years come together too. Good *years* promise even more than the good *year.* The inevitable long string of bad years is even more catastrophic.

If this is starting to sound familiar, then here's the exact citation: Genesis 41, starting at verse 14. Egypt's Pharaoh is having bad dreams about cattle and ears of grain, seven at a time. Joseph interprets them: "There will come seven years of great plenty throughout all the land of Egypt, but after them will arise seven years of famine, and all the plenty will be forgotten in the land of Egypt; the famine will consume the land."

Over 2000 years of observations suggested to an English scientist, Edwin Hurst, that the Nile hasn't changed since Joseph's time. The Nile, and particularly how much water it has at its lowest ebb, determines feast or famine. The Nile integrates the rainfall across the drylands of Africa from equatorial Lake Victoria northward to the Mediterranean. What Hurst discovered from analyzing the records of the Nile's lowest ebb is what Joseph also knew. Dryland rainfall varies from year to year, but with runs of good and bad years. There is nothing special about the number 7—the runs are of various lengths. But there is a tendency for the good to come with the good and the bad with the bad. (Hurst is known for his index that characterizes that tendency.[24])

It is hard to imagine a more malevolent combination. Unlike the repeated throws of dice, good and bad years do not occur at random. A series of good years lure in the unwary, and then the longer periods of bad years ruin them. Like the most addicted gambler, the farmer will not leave the game. Surely, next year will be a wet one! Unlike the gambler, the farmer or grazier mesmerized by the good years and beaten by the bad ones destroys that land when he or she stays in the game too long. For the world's poor, it's the only game in town.

In *Bad Land*, Jonathan Raban recounts the history of his family's neighbors on the Montana prairies early in the last century.[25] In Montana, the lure was a railroad to get them there, the Homestead Act—a free 320 acres of land (over a square kilometer)—and Hardy W. Campbell's *Soil Culture Manual*, placed next to the family Bible, to tell them how to farm it. They built their two-story farmhouse and barns and created a "brave and hopeful civilization on the prairie." In the 1930s, after little more than a decade, they abandoned them intact.

Sixty years later, in the hard-to-find traces, Raban found an old bonfire and, within it, a stark reminder of failure. When they left, the family wanted no memories; they burned their family photos before leaving. Just one reminiscence is passed down, a single sentence by the settler's son. He "hoped he'd never again have to see anything like his mother, down on her knees, day after day, praying for rain."

In his article, "The Dirty Thirties," University of Kansas historian Donald Worster also explores how and why we didn't adjust to the vagaries of nature during the 1930s.[26] International events put the American wheat farmer in an enviable position during World War I. The Turks cut off the shipments of grain from Russia, then the largest wheat exporter in the world. From 1914 to 1919, Kansas, Colorado, Nebraska, Oklahoma, and Texas expanded their wheat land, plowing up 44,000 km^2 of native grassland to do so. "Wheat will win the war!" read the posters. Many Plains farmers made fortunes in the "Great Plow-up." The mutualism of bankers, farmers, and tractor manufacturers kept up the pressure. By the time the dust began to roll across the plains in 1935, 130,000 km^2 of former grassland lay naked and vulnerable to the winds.

Worster explains that it was due to neither ignorance nor lack of experience. Cattlemen had driven their stock to railheads across these plains, and farmers had broken the sod, leaving a record of devastation as well as years of success. There was some understanding about the land's severity. Those who plowed it included the well-educated, people with access to capital and expertise. None of the warnings, the experience, or the knowledge stopped the destruction that was to take place. The risk-taking economic culture, at fever pitch in the 1920s, was impervious to more cautious attitudes toward economic expansion and environmental change.

In Africa, it is not short-term greed that lures people onto the marginal drylands; it is population pressure. The malevolent sequence of a few good years draws in people desperate to eat. In the bad years that follow, they and their animals starve and destroy the vegetation, perhaps irreparably.

Civil strife complicates the ecological story of unpredictable and often severe droughts. The most recent one lasted almost 20 years and, as a result, Lake Chad contracted at its low point to one-third of its normal area. Scenes of fighting and starvation in African countries are an obvious intrusion even into American TV news, with its inability to grasp stories about peoples and its predilection for stories about personalities.

It's too much of a coincidence that a map of African countries that suffer prolonged civil strife looks like a map of dry, marginal lands. Across the Sahara's southern edge a belt of drylands stretches from the Atlantic to the Indian Oceans. In the west, Mauritania, Mali, Burkina Faso, Niger, and Chad make up the region known as the Sahel. The drylands continue eastward without a break across Sudan, Ethiopia, northern Kenya, and Somalia. All have fighting and famine. As Norman Myers and Jennifer Kent have so carefully documented, all have many millions of "environmental refugees"—people who flee their homelands because of drought, soil erosion, desertification, and other environmental problems.[27]

Will the Land Recover from Our Abuse?

On the cooled black lava flows of Hawai'i, bright green ferns nestle in sheltered cracks. As the lava ages and crumbles, organic debris accumulates, and other plants join them. Given enough time—decades or even centuries—a lush forest will grow. If life can establish here, surely it will reclaim the lands that we have damaged.

The question is not "Will the land recover?" but "How long will it take?" How long will soils with high salt concentrations remain barren? How long will it take soil to reaccumulate in places where it has washed away? How long will it take for grasses to reclaim areas now invaded by drought-tolerant shrubs? How long will it take for areas stripped of all woody plants for firewood to recover?

The answers to these ecological questions are not encouraging. At best, a couple of decades may be enough to allow abandoned forest to recover its fertility and thus be worth clearing again. There may be soil 100 years from now on islands formed of fresh volcanic ash. It generally takes 100 to 300 years to form just 1 cm of soil, hence thousands of years to get enough soil to form productive land. Forests may take several hundred years to recover.[28]

Lands lost to salt accumulation or shrub invasion yield few answers, for we are not yet successful at reversing either of these processes. The shrub invasions and arroyo cutting in the Southwest are still evident a century later. Indeed, abandoned lands are enough in evidence to yield a simple reply to these questions. We do not know how long many lands will take to recover, for we have not waited long enough.

Journey's End

This chapter ends our journey to assess how well we are managing the portfolio of biomass-generating accounts. It has contributed pieces much like the two that preceded it. Also as in Chapters 3 and 4, it has provided insights into human impact that cannot be encapsulated into a simple amount of biological interest withdrawn.

Perhaps this chapter's most important point is that no single number can capture what we do to the planet. There are good years and bad years.

We may take two-fifths of the land's production each year on average, but some years will be better and some will be worse. We still have to live through the worst years. Prudent managers of financial investments know to set resources aside for "rainy days." What a silly image: It is the dry years we really have to worry about!

Those who broke the sod of America's prairies early last century knew better. The drought arrived with the certainty Joseph had attributed to God's will. Restraint and the foresight to protect land from rainless days came second to short-term greed. Plow in haste, repent at leisure. Those who destroy the Sahel may know better too, though there is precious little they can do about it. Those who chopped the last palm on Easter Island to build a seagoing boat may have understood their actions too. They still died when their land died. And Easter Island has remained treeless to this day.

6

"Man Eats Planet!
Two-Fifths Already Gone!"

It is time to do the accounts and produce the balance sheets. Yes, I know that this sounds like doing one's taxes or balancing a checkbook: It is painful but nonetheless unavoidable. This chapter is also by far the shortest—and its bottom line is only a few pages away.

To arrive at that bottom line the Stanford group first calculated the total interest—how much plant biomass the land surface produces each year. In Chapter 1, I cross-checked their estimate of 132 billion tons and thought it was about right, if perhaps a little on the high side. In any case, it is a number that must vary from year to year.

The second step of the Stanford group was to calculate how much of this production we withdraw. They did so in two ways. The first was the answer to the "What-does-the-planet-do-for-you?" question. Chapter 2 counted up three numbers that go into this first question: the plant growth we consume (item 1 in my spreadsheet), the plant growth our animals consume (item 2), and the wood we use (item 3). The total was 5 billion tons. This is only a small fraction of the total annual production.

It is the answer to the "What-we-do-to-the-planet?" question—what we do to get our food, our animals' food, fuel and wood—that is much larger. The Stanford group recognized two kinds of numbers that contribute to this much larger total. The first kind involves how much of the land's current production goes to support human activities. An example of this was the plant production across all the world's croplands—leaves, stalks, and other inedible parts—that total much more than the small parts of the plants we eat. These are the ways in which we use plant production either through direct or indirect consumption.

The second set of numbers adds in the amounts of production that we forgo because of those activities. The first example (item 5) was the conversion of productive land to cities, roads, and similar uses, none of which produces biomass. (This two-part logic applies to your investments. You might proudly announce that you are using only 90 percent of the annual interest that they generate. Now suppose that this interest is only half what it would have been if you had not spent half your principal on a new car last year, in which case you forgo half of the interest.) These numbers involve urban areas, croplands, forests, and grazing lands. I will consider them in that order.

Urban Areas

Item 5 in my spreadsheet was the estimate that our cities and roads cover about 2 million km² of what was once productive land. These areas are often in places that would have been very productive land prior to our settling them. The Stanford group used an estimate of 1500 tons of biomass per square kilometer for the production of this land were it not urbanized. I agree and write down the first subtotal.

> We forgo 3 billion tons of plant production each year as a consequence of needing places to live.

Croplands

The Stanford group considered two numbers for croplands. The first was the total production we use across the 15 million km² of croplands. The second was how much production we might forgo as a consequence of converting this land from its naturally more productive state.

The area of croplands is not controversial. It is one of those numbers we should expect to know well, for our very lives depend on how much food we grow. There is controversy on how productive cropland is and how productive the land was originally.[1]

The Stanford group used a lower overall estimate of how many tons of biomass are produced per square kilometer than I did. Then they worried that this plant production per square kilometer may be too low a number, admitting that the production of croplands was the number in which they had least confidence. Higher values of plant production on cropland appear sensible in light of the enormous advantage that crops enjoy over natural ecosystems: fertilizer.

Apply nitrogen fertilizer to your lawn and it quickly gets lush, green, and more productive. That is why you have to mow it more often. The same principle applies to crops. And the amount of fertilizer we produce is staggering.[2] Industrial production, intended for the 15 million km^2 of crops, now exceeds the total production of all nitrogen by natural ecosystems on the remaining 115 million km^2. Compilations of the hundreds of studies of plant production done since the Stanford group published its study suggests that croplands have an average production of 1700 tons per square kilometer. If we multiply this by the area of croplands, we derive the first big number.

We use 26 billion tons of biomass each year from croplands.

How much production do we forgo because croplands are less productive than the natural systems they replaced? Desertification causes the land to forgo plant production. It destroys a few tens of thousands of square kilometers of irrigated and rain-fed croplands each year (Items 12 and 13). In addition, about 2.5 million km^2 of these croplands have less production than they might otherwise enjoy (Items 12 and 13).

Dramatic though the lost fields are—especially if they are your own—the total area is relatively small. And how much is the reduction in plant production across the degraded 2.5 million km^2? The losses would have to exceed a quarter of the average production of croplands before more than a billion tons of potential production would have been lost. The Stanford group ignored both these losses, deeming them too small to contribute much to the grand total. They were probably right to do so.

Forests

The use of forests is a sum of two subtotals, and each of these has several parts. The first subtotal involves the wood we consume. Tree plantations are the forestry analogue of agriculture, so we should count all their production, or about 3 billion tons each year (Item 8). In addition, there is the wood we harvest from temperate, boreal, and tropical forests. That extraction is large enough to keep the temperate and boreal forests dominated by young trees rather than the mature forests they would become naturally. Item 7 estimated the amount from the first two forests alone at 0.75 billion tons. Cutting tropical forests would add substantially to this amount.

Then there is the wood burned for fuel, mostly in Africa and Asia, and thus in dryer woodlands than the ones counted in previous totals. That was estimated at another billion tons (Item 3).

Combined, these numbers come to nearly 4.75 billion tons before we add in the cutting of tropical forests. Five billion tons does not seem an unreasonable guess under the circumstances. That is exactly the number the Stanford group estimated, though they obtained their result a slightly different way.

> *Subtotal: 5 billion tons of production go to support our timber extraction, wood burning for fuel, and associated uses each year.*

The second subtotal of the use of forests is a large number that is much more difficult to estimate. It is the wood we destroy to get things other than wood. Tropical forests shrink by 160,000 km^2 each year (Item 9). An area perhaps several times that size is burned or severely damaged each year (Item 10). Vitousek and his colleagues estimate that the permanent losses of forests are equivalent to a loss of about 3 billion tons of production each year. The burning and transient uses are equivalent to the loss of over 6 billion tons of biomass, again each year. They certainly seem to have the right ratio; much larger areas are burned than cleared. But how did they derive these answers? They are in terms of the loss of annual production, not in terms of the area of forests cleared or damaged.

The Stanford group estimated these two numbers by considering the movement of biomass from the land to the atmosphere and back, a process that requires a little explanation. The biomass that has preoccupied the previous five chapters is mostly carbon. We burn biomass and, in doing so, produce carbon dioxide, a gas that permeates the atmosphere. The carbon dioxide concentration of the atmosphere is measured precisely, year after year, and from pole to equator to pole. It is on the verge of doubling from what it was a century ago. Our burning of fossil fuels, forests, and grasslands is the cause of this doubling. Now, because carbon dioxide is a greenhouse gas, higher concentrations in the atmosphere increase the amounts of energy that remain near the planet's surface. The increasing concentration has uncertain consequences for the planet's temperature; therefore, information about how much carbon goes where, and why, and how is of crucial concern to a broad spectrum of scientists.

The larger share of the growing carbon dioxide concentration is because we burn fossil fuels. The smaller but substantial share of the increase is because we clear and burn forests. These numbers—3 and 6 billion tons, respectively—are calculated from how much carbon dioxide is contributed to the atmosphere by the processes of clearing and burning. Do these numbers make sense with respect to the area of forest cleared? I think they do, but that alone does not justify the numbers. I have added a brief explanation to the technical notes at the book's end.[3] Together, these amounts give the second subtotal:

> *Subtotal: 9 billion tons of production go into the atmosphere each year as a consequence of forest clearing and burning.*

Some of the 9 billion tons that are burned or cleared are also used for fuel or harvested (estimated to be 5 billion tons), but the amount of overlap is likely to be quite small. So combining the two subtotals gives the next big number.

We use 14 billion tons of forest production each year.

Grazing

The numbers for grazing also come in several pieces. About half of our domestic livestock are on natural grazing lands; the other half are on converted forests. For the half on natural grazing lands, their direct consumption should be half of the 2 billion tons estimated for all livestock combined (Item 2). In addition, we deliberately burn these natural grazing lands to enrich them. Vitousek and his colleagues estimated that this burning covers 6 million km^2 (Item 14) each year and accounts for another 1 billion tons of biomass. They obtained this amount in the same way as for the burning and clearing of tropical forests: by how much burning raises the carbon dioxide level in the atmosphere.

> *Subtotal: Grazing on natural grazing lands plus their enrichment by burning uses 2 billion tons of biomass each year.*

The second half of the calculation adds in the livestock on pastures that were converted from other kinds of ecosystems. Altogether, about 7 million km^2 of land that was once forest have now been converted to permanent pasture. This includes 2 million km^2 from temperate forests (Item 6). By and large, this conversion has produced the productive green fields of Europe, eastern North America, New Zealand, and elsewhere. The rest is the 5 million km^2 converted from tropical forests (Item 11). Tropical forests have shrunk by 7 million km^2 (Chapter 4), but only 2 million km^2 have become croplands. The rest is nominally grazing lands, though much of it is pretty pitiful.

The Stanford group estimates that the total areas of 7 million km^2 would have produced about 1600 tons of biomass per square kilometer each year for a total of just over 11 billion tons. I think those numbers are sensible, given what we know about forest productivity.

How much this land produces now that it is grazed is more controversial. The Stanford group estimates that this land now produces 1400 tons of biomass per square kilometer each year. So the amount we forgo at 1.4 billion tons is a paltry loss of 200 tons per square kilometer.

I do not believe the loss is that small; I think we forgo far more biomass than this. Across the 7 million km^2 of "once forest, now pastures," my opinion is that the land is almost certainly less productive than it once was. Over many areas of cleared tropical forests the land looks like a wasteland, with a few scrawny animals. For instance, FAO's numbers show that the Democratic Republic of the Congo (Zaire) has slightly more permanent pasture than does New Zealand. Those pastures support only a tenth of the domestic animals. In other countries, some of the converted forests are pasture to only the most optimistic government bureaucrat; they are on rapidly eroding hillsides.

This disagreement is about how much production we use and how much we forgo. As long as we add the pieces, it doesn't make any difference.

Subtotal. In converting forests to grazing land, our activities use and forgo a total of 11 billion tons of biomass each year.

The final number is the extent to which grazing drylands produce less because of the damage we have done to them. The Stanford group estimated that desertification has reduced plant production across 35 million km^2. That is the same number as Dregne's (Item 15). Desertification is severe on 15 million km^2. They take those 15 million and assume that severe desertification means a loss of a quarter of the plant production. This comes to a reduction of about 4.5 billion tons, which smacks of brave but plausible guesswork. I round it up to 5 billion tons to add in the loss of production from areas suffering only mild to moderate desertification.

Subtotal: We forgo 5 billion tons of biomass each year by our grazing animals on drylands

Add the three subtotals.

We use and forgo 17 billion tons of biomass per year for grazing our domestic animals.

The Bottom Line

Each year, we use 26 billion tons of plant production for crops, 14 billion tons from forests, use and forgo 17 billion tons for grazing, and forgo 3 billion tons for urban areas. That comes to a total of 60 billion tons. Chapter 1 showed that the land produced 132 billion tons even after these activities. Were we not to have cities, roads, desertification, and conversion of productive forests to poor grazing lands, the land would produce even more. The Stanford group thought about 10 billion tons more, for a total of 141 billion tons. The 60 tons from the 141 gives us—yes, you guessed it—42 percent.

That's it. We are done.

Arguments about numbers make less difference to the bottom line of human impact than you might think. Chapter 1 estimated the total production at 115 billion tons. Add in the 9 billion that we forgo, and the total is 124 billion. If this lower number were correct for the land's production, then the previously summed human impacts would constitute 48 percent of the potential production of the land.

One can tinker with these numbers in lots of other ways too, choosing one number over the next and seeing what effect it has. You may be relieved to know that I have no intention to do that here. However one cuts the numbers, their broad quantitative conclusion emerges: We already take about two-fifths of the land's production.

Was there an easier way to get two-fifths without the details? How wild a guess was my "42 percent" in reply to Ehrlich's question? Of the approximately 150 million km^2 of the planet's land surface, only about 100 million km^2 contribute much to the total productivity. The rest is too cold or too dry. Of this useful land, 15 percent is cropland, at least 30 percent is pasture of different kinds. From these uses alone we co-opt about 45 percent of the usable land surface on a per-area basis.

That was my first stab at the answer. Then I thought some more. We also occupy the most productive lands first and avoid the less productive lands. So the 45 percent of the land area we use is likely to be more productive than the 55 percent of the area we do not use; a figure of 45 percent is therefore too low.

Yet we will not use anywhere near all the production on pastures. Most pastures include large areas where wild animals graze along with sheep, cattle, and goats. So the withdrawal should be less than 45 percent. Conversely, we also use forests, so the number might be higher than 45 percent.

"If it is that easy, why did you take a third of a book to explain it?" I hear you ask in desperation. As with other journeys, the trip may be more interesting than the destination itself. It is the pieces of the calculation that are so important.

What the Pieces Mean

What we appropriate for agriculture is the largest piece: 26 billion tons from the total of 60 billion. The area of cropland that produces this tonnage is not increasing by much. In the 30 years since 1970, a period when the human population nearly doubled, the area of crops grew from less than 14 million to slightly more than 15 million km^2. Agriculture feeds far more people from almost the same area of land. In this statistic, eternally optimistic Panglossians see continuing progress that allows us to push the envelope of our environmental constraints. In the same statistic, worried Cassandras notice the constraints on the area of croplands that prevent it from expanding with our growing population.

Some of the smaller pieces that contribute to the remaining 34 billion tons are productive areas in permanent pastures converted from temperate forests, forest plantations, and urban areas. In total these occupy about 5 million km^2 (approximately 3 percent of the land's surface), and they account for 5 percent of the land's production. These areas are also not changing quickly. Certainly plantations are growing rapidly on a proportional basis—about 5 percent of their area per annum—but that area is still less than 1 percent of the land surface.

Again, one may view these numbers with considerable optimism. A greater percentage of people are living in cities and thus sparing the land that might otherwise become suburbs. Even the increase in plantations appears to be a good sign. The more wood we grow there, the less we will need to extract from natural forests. If the story ended here, then perhaps we should have Pangloss's invincible enthusiasm for all we see. But it does not end there.

The 6 billion of us are already inflicting continuing damage. The larger pieces of the remaining 34 billion tons include the massive shrinkage of tropical forests and the large amounts of production that we forgo due to damage done to tropical forests and drylands. Moreover, we must interpret the story with care if we are to make projections about how the impacts will change in time.

Developing nations increase in population; their citizens aspire to the same standards as the ones in developed nations (grain-fed beef and the rest of it). Countries with established market economies continue to demand ever more resources from these nations. Our withdrawals of biological interest from the planet's savings account will increase, and they will do so faster than the rise in human population.

The richer nations have taken the most productive land; that is an important reason for why they became rich. The developing nations have the poorer land; that is why they have not developed so quickly. As the last two chapters have shown, from deserts to rainforests the conversion of poor land to crops and pastures does not usually mimic the fate of converting the best land. We damage the poorer land far more in the process. Yet it is from the poorer land that the increasing withdrawals must come.

In a metaphorical sense, Gaia, our hostess at the global cocktail party, is not happy. There are always people who high-grade the cashews from bowls of mixed nuts. Not for them the lowly sunflower seeds, still in their husks. Now this is all that remains. Those who did not get the cashews are asking for them. The high-graders who did get them are now demanding the even more expensive Macadamia nuts. The doorbell is ringing, announcing the arrival of more guests. Dr. Pangloss is telling her in one ear that everything will be OK. In her other ear Cassandra is assuring her that everything is not OK. What Gaia knows for certain is that in cracking the sunflower seeds, the guests will make one hell of a mess, drop bunches of them on the carpet, and spoil the party for everyone.

7

Water, Water Everywhere?

In KwaZulu-Natal, South Africa, the women and children gossip and chatter as we stop to talk. Gathered around the pump they have every reason to be happy. With the financial help of the nearby mining operation, they have already achieved President Mandela's goal for his country: a source of drinkable water within 200 m of their homes.[1]

Laughing, they gesture to our wives, indicating they have found good husbands. My colleague, Rudi van Aarde, tells me this translates to our showing the results of too much food and too little exercise. Not a problem for the ladies by the pump; the luckiest of them load the 25-liter (25-L) water-filled yellow plastic containers onto wheelbarrows and trundle them home. The less fortunate perch these 25-kg weights on their heads, balancing them with enough skill to free one hand. The yellow containers, labeled "cooking oil," are everywhere in Africa. Do the people who manufacture them know of their essential contribution to African village life?

President Mandela's goals are admirable and seem modest. In fact, they are extremely ambitious. Where is the water to come from? Two-thirds of South Africa receives less than 0.5 m of rain each year. South Africa's population is growing: It will double in the next 30 years.[2]

Are we running out of water? Listen to political leaders worldwide. They answer "yes," and do so with an unfettered anger toward those with whom they might share that water.

Water Wars

Look at the conflict in the Jordan River Basin. King Hussein declared in 1990 that water was the only issue that would take him to war with Israel.

The rhetoric comes from both sides: Former Israeli agriculture minister Ben-Meir said much the same thing.[3] It's not hard to see why. Even with their current populations, Israel, Jordan, and Syria are net importers of cereals. All three countries have explosive populations. Israel is trying to house a million immigrants from the former Soviet Union. Jordan's population will double in 20 years and Syria's in 18.[4]

Israel and Jordan have similar amounts of cropland (about 4000 km^2) and similar populations (about 5.5 million). Jordan uses only half as much water and irrigates 16 percent of its croplands, compared to Israel's 42 percent.[5] Jordan's expectations of greater affluence would greatly impact its water use, even if its population were not increasing; Israel's consumption already exceeds its supply of renewable freshwater.

The Palestinians' quest for an independent West Bank homeland would be difficult enough were it just a matter of the land, language, and religion. Some 25 to 40 percent of Israel's sustainable freshwater supply comes from an aquifer that lies beneath the West Bank—the land Israel occupied after the 1967 war. During its occupation, Israel has severely restricted the amount of water that the West Bank Arabs can pump. Yet Israel has overdrawn the aquifer for its own use.[6]

Israel also occupies the Golan Heights, a part of Syria. Control over this area gives Israel access to the slopes that drain water into the Yarmuk River. This is the last undeveloped river that flows into the Sea of Galilee, and it forms the boundary between Syria and Jordan. Galilee is Israel's largest source of surface water, and from it a large canal and pipeline, the Kinneret-Negev conduit, take the water far to the south. Jordan and Syria have plans to dam the Yarmuk at Maqarin. Israel has announced that, if built, they will destroy it, fearing that it will reduce the amount of water Israel can extract from Galilee. The bellicose Golan Residents Committee, complete with English-language brochures and an email address, makes its position on this occupied land clear: "There is no room for territorial compromise . . . the Golan controls 30 percent of Israel's water resources."[7]

Israel's use of fertilizer is seven times that of Jordan's—another statistic that Jordan would like to equalize.[8] Yet there are consequences: The water that flows back into the Jordan River after use by Israel's agriculture is already extremely polluted by fertilizers. There is industrial and urban pollution too. And the heavy pumping of the coastal aquifer has led to seawater seeping into it.

Is this an isolated problem? Hardly! Sandra Postel, a scientist at the Global Water Policy Project in Massachusetts, discusses these and other areas of conflict in her excellent and highly readable book *Last Oasis: Fac-*

ing Water Scarcity.[9] "India and Bangladesh haggle over the Ganges River," she writes, "Mexico and the United States over the Colorado, Czechoslovakia and Hungary over the Danube, and Thailand and Vietnam over the Mekong." Eight other countries could control the flow of the Nile into Egypt. Like Egypt, all have high population growth. Several are among those countries that the FAO considers as having "low food security." None is famous for cooperating with others.

Like the green stuff of the previous chapters, water is another common currency, and it is one that we can convert into plant production. This chapter asks the same two questions about water as the previous chapters did about plant production. The first is the "What-does-the-planet-do-for-us" question. How much water does the planet supply us, and what fraction of the world's annual supply do we use? Half the ice-free land surface is too dry to grow much food unless it is irrigated. So why don't we irrigate more land? As the preceding examples demonstrate, water is in such short supply that some countries threaten war to get it. The second is the "What-do-we-do-to-the-planet?" question. What do we do to our rivers, lakes, and wetlands to get the water we use?

Our Need for Water

How much water do we need? Peter Gleick, a scientist at the Pacific Institute for Studies in Development, Environment, and Security in Oakland, California, arrives at a figure of 50 liters per person per day. That's just for personal use—not the extra amount for growing food or for industry. That comes to 18 tons per person per year.[10]

Only 10 percent of that 50-liter allowance is needed for drinking in hot climates, much less in cooler ones. For cooking and kitchen uses, Gleick suggests another 10 liters, for bathing another 15, and a final 20 for flushing away human waste. He goes on to suggest that 50 liters should be a basic water requirement, a bare minimum that should be considered a human right, one established first and not one to be negotiated downward in talks of who gets what water.

For comparison, the average Californian uses over 10 times this daily allowance, flushing more than 120 liters down the toilet, 100 in the shower, 130 in the laundry and the kitchen, and nearly 200 making the grass green and the garden grow. The figures for The Netherlands and Sweden, among the more environmentally conscious countries, are 100 to 200 liters per person per day.

About 1 billion people live in countries where the average water use is less than the 50-liter amount. Most of them are in Africa. Many more people live with less water in countries such as South Africa, where some have much more than 50 liters and others much less. China and India, for example, have average domestic usage only slightly over 50 liters: At best, nearly half of the people in these populous countries lack the basic water requirement.

For Israel, Jordan, and the West Bank, the combined 12 million people need more than 200 million tons of water to meet this basic requirement now, and will need twice that 20 years from now. The total supply comes to about 3.5 billion tons, which suggests there is a lot to spare.

So why is there a conflict? It is crops that need so much water. Postel is also the lead author of a paper that appeared in *Science* in 1996 that attempts a global assessment of how much water we use for drinking, agriculture, and other uses and how much water is accessible.[11] She is joined by the now-familiar Stanford University firm of global accountants, Ehrlich and Associates, and their new junior partner Gretchen Daily.

The word *accessible* is the key. Freshwater constitutes only about 2.5 percent of the total volume of the Earth's water. "Water, water everywhere, nor any drop to drink . . ." The ancient mariner of Coleridge's poem proclaims his woe at being surrounded by water, but it is the sea and he cannot drink it. His lament is a fitting metaphor. Our crops and our livestock are no more able to use seawater than is the sailor. As I shall now explain, there is a lot of freshwater that we cannot use either.

We can remove the salt from the seawater, but not economically so. As a result of its high cost, desalination supplies only about one-thousandth of the world's freshwater. At present, it is used to produce drinking water for the water-scarce but oil-rich countries around the Persian Gulf. Two-thirds of the freshwater is frozen in glaciers and ice caps. Towing icebergs is an option unlikely to produce large quantities in the foreseeable future.

The rest of the freshwater amounts to about 11 million km^3. It is held in aquifers, lakes, and rivers, and, most important of all, in the atmosphere. The largest fraction of this is the rain that falls over the oceans from where the water evaporated in the first place. Only one-hundredth of the freshwater that is not frozen falls over the land each year. It is the use of this water that Postel and her colleagues investigate.

A total of 110 trillion tons of rain falls over the land, another big number that has one wondering whether or not it makes any sense. It is equivalent to 0.85 m of rainfall evenly spread across the land surface each year. That seems to be about the right number; it corresponds to a modest amount of rainfall, neither a desert nor a rainforest. In any case, there is good agreement on this number. Experts' estimates of total rainfall do not vary by much: 90 trillion is a low estimate and 120 trillion is a high one.

There are two pathways for this water. The larger one, about 70 trillion tons, is called *evapotranspiration*. What evaporates either directly from the ground or is transpired through plants (soil to roots to stems to leaves to air) usually equals or exceeds the rainfall. It is hard to estimate separately the amounts of water evaporated and transpired, so ecologists combine them.

Another roughly 40 trillion tons run into rivers and back to the sea. Called *runoff*, this is not so easy to estimate, because the total is reached by summing the flows to the ocean of the world's rivers, which include a few large ones and many, many smaller ones. In order of flows, the continents are ranked with Asia first (15 trillion tons), then South America (10), North and Central America (6), Africa (4), and Europe (3); Australia is last (2).

Water for Plant Growth

We use water along these two pathways in very different ways, and so Postel and her colleagues tackled them in turn. The 70 trillion tons of evapotranspiration provide the water for plant growth, which Chapter 1 tells us totals roughly 130 billion tons. Of this total plant growth, we use about 50 billion tons each year, and we likely use the corresponding fraction of evapotranspiration.

An even rougher estimate would be that since our croplands produce 26 billion tons a year, they need water in the same proportion—or about 14 trillion tons. Postel and her colleagues show that these rough calculations are on the high side. They estimate that croplands need only about 6 trillion tons of water, suggesting that we have developed crops to grow in drier than average places. There is a way to check this. Six trillion tons over 15 million km^2 means that crops use an average of six-fifteenths (0.4) m of rain each year. Roughly a third of all rainfall runs off the land worldwide: 40 trillion out of 110 trillion. So for crops to have 0.4 m to use, it has to rain 0.6 m each year.

That does seem about right. In England, Europe, and Russia the cities of York, Paris, and Moscow sit among rich agricultural areas, and each gets slightly more than 0.6 m of rain in an average year. Eastern North America gets a bit more: Des Moines, in the middle of Iowa, gets about 0.8 m.[12]

As you would expect, agricultural scientists know a good deal about how much water it takes to grow a crop. Some crops are thirsty; rice needs the equivalent of nearly 2 m of rain and produces more when flooded. Corn needs about 1 m, but other crops need less. Wheat, sorghum, and millet can make do with the equivalent of 0.2 to 0.25 m of rain per year, though under these arid conditions the grain yields are low.

There is an important way to look at these numbers: It takes 6 trillion tons of water to feed 6 billion people. Each one of us needs an average of 1000 tons of water per year to grow the food we eat. Armed with these numbers, the political representatives of agriculture are understandably belligerent in the Middle East (and elsewhere). The current population of Israel, Jordan, and the West Bank (12 million) would need 12 billion tons of water each year to grow their food, not the 3.5 billion that they have. And their industries have to consume water as well. By 2025, when the UN's estimate of the combined population size reaches 22.3 million people, the shortfall will be even worse.

Water is in desperately short supply in some parts of the world. In such places the answer to the "What-does-the-planet-do-for-us?" question is a strident "Not enough!" Hear that reply! You can be certain that the answers to the "What-do-we-do-to-the-planet?" question include drastic changes to the natural landscape.

Runoff

The second pathway for water is the *runoff*. The roughly 40 trillion tons of water that run off the land are six times the amount needed to grow our crops. Can't we use more of it to grow the food our rising population requires? The problem is that a large part of it is in the wrong place or flows at the wrong time to be useful to us. Some water is both: It runs off in floods from geographically remote places.

August is a dry month in Manaus on the Amazon. The river's water has taken months to get here from the Andes, and it is now at its highest level. Looking across from outside the fish market, it is hard to accept the fact that the river is still 1000 km from the ocean. It does not even look like a river. There is little sign of movement and the low jungle that forms the other bank is on the far horizon. The guide starts up the canoe's outboard and we skim over the smooth water. We turn right up the Río Negro, and, after a couple of hours, head up a large tributary, then smaller ones until the forest closes in. The water's surface is a mirror that reflects the trees so perfectly it is hard to tell which way is up. More magic, for we travel through the very tops of these trees. This is the *igapo*, the flooded forest. The forest floor is 10 m below us, under water. There the trees drop their fruit to be eaten (and their seeds to be dispersed) by fish.

By December, the rains are daily and the river has dropped. This time I go left, cross the clear Río Negro past where it joins the muddy Río Solimões, mingling like cream poured into coffee, and to the south bank.

Dolphins ride the boat's bow wave. Tall islands of gray mud rise above the water; they were completely submerged four months earlier.

Some 6 trillion tons of water flow down the Amazon each year: 15 percent of the global total. As benchmarks, the Mississippi's annual flow is about half a trillion, the Danube, a fifth of a trillion—one-tenth and one-thirtieth of the Amazon, respectively. In the right place, the Amazon's water would be enough to grow crops for another 6 billion people each year.

It is not in the right place. Only 25 million people live in the Amazon basin; that is 0.5 percent of the world's population. South America has only about 5 percent of the world's population, but its rivers, including the Amazon, have over a quarter of the total runoff.

The Amazon is not the only river in the wrong place. In Africa, the Congo contributes over a trillion tons (3.5 percent of the global runoff), yet its basin supports slightly more than 1 percent of the global population. The Lena (0.5 trillion tons), Ob (0.4 trillion tons), and Yenesey (0.6 trillion tons) flow into the Arctic Ocean, draining sparsely populated Siberia and parts of Kazakstan and Mongolia. These smaller rivers and their counterparts in sparsely populated Alaska and Canada add nearly another 2 trillion tons to the total runoff. Not all of this water is in the wrong place, for people do live along these rivers and do use some of it. Nonetheless, Postel and her colleagues estimate that a total of perhaps 8 trillion tons of runoff is inaccessible.

The Amazon's flow does not come at the right time either. The seasonal flood I have described means that the water cannot be distributed throughout the year over areas that might grow food. It is not unique. Most runoff comes in floods. The three Siberian rivers flood in the springtime as snow and ice melt. Worldwide, about three-quarters of the runoff comes as flood water. So, some of the global runoff flows in the wrong place, some comes at the wrong time, and some, about 6 trillion tons, does both.

Dams can capture floodwater that is in the right place. In practice, they get about 12 percent of it, some 3.5 trillion tons. Postel and her colleagues add this to the 9 trillion tons that do flow at the right time and in the right place to get a global total of accessible runoff of 12.5 trillion tons. Postel's group shows that there is runoff that we consume—it is not available for any other purpose—and there is runoff that we borrow and return to the rivers.

Who Uses the Runoff?

Agriculture takes the major share of the accessible runoff. There are about 2.5 million km^2 of irrigated agriculture worldwide. Another very rough estimate: Recall that, on average, crops need 0.6 m of rain to grow. That depth, averaged over the irrigated land, would amount to 1.5 trillion tons of irrigation water. That is surely too low a value for agriculture's share. The countries that irrigate more than half of their cropland grow thirsty crops such as rice (Bangladesh, the two Koreas, Japan, Pakistan, and Surinam) or are hot and dry (Egypt, Iran, and the central Asian republics of Azerbaijan, Georgia, Kyrgyz, Tajikistan, Turkmenistan, and Uzbekistan).

Not surprisingly, Postel and her colleagues suggest a higher value: Agriculture uses about 2 trillion tons of runoff. It also borrows another trillion tons and returns it to the rivers, for a total consumption of nearly 3 trillion tons.

Worldwide, industries use much less water than agriculture but borrow much more, often returning it, the worse for wear, in a polluted state. So, too, do cities, whose inhabitants will use much more water for flushing away waste than they will drink. The reservoirs that store the accessible runoff also use some if it; that is, it evaporates from their surface. Combined, these activities consume almost half a trillion tons and borrow another 4 trillion.

In total, we use and borrow 17 percent—about one sixth—of all the global runoff. As a fraction of the 12.5 trillion tons of accessible runoff, our use is much higher—nearly 60 percent.

I hope previous chapters inculcated a sensitivity to such a high percentage. It is half as large again as the fraction of plant growth we withdraw from the land each year. While remote rivers such as the Amazon and the Congo are almost untouched, a percentage that high for the others means that we probably make massive impacts on them.

What Do We Do to Our Freshwater Ecosystems?

What are these impacts on the rivers and the ecosystems that they drain? And how do we measure those impacts? What currency do we use? There is no easy answer. And I know of no global answer at all, though Christer Nilsson, from Umeå University in northern Sweden is working energetically to provide one.[13]

His first contribution, written with Mats Dynesius, assembled a mass of information about the largest rivers in the northern third of the world.[14] Together these rivers account for about one-fifth of the global runoff and a larger share of the accessible runoff. They published the results of their heroic efforts in *Science* in 1994.

Seventy-four of the rivers were in Canada and the United States, and an additional 139 were in Europe and the former Soviet Union. Their numbers were not simply to be found in convenient global summaries. Some did come from UN documents—*Discharge of Selected Rivers of the World*, volumes 1 to 3—great reading for Umeå's long winter nights, I assume. Other numbers came from obscure technical reports, written in a dozen different languages, one country or sometimes one river at a time, from the Amu-Dar'ya in Afghanistan to the Yukon in Alaska. They requested yet other information directly from the local or national authorities that manage the rivers: the Office of the Delaware River Master or the State Hydrological Institute in St. Petersburg, Russia.

Dynesius and Nilsson presented four sets of numbers: (1) the river's rate of flow, (2) its fragmentation by dams, (3) the capacity of its reservoirs relative to how much water flows down the river each year, and (4) how much of the river's annual flow is used by irrigation. All are different measures of how we impact rivers. But even combined they are incomplete. None measures how badly polluted a river is, for instance.

Wild Rivers, Tamed Rivers

The term *fragmentation* needs to be explained. Dams fragment the original river and break it into unaffected river segments separated by flooded stretches. *Fragmentation* is the fraction of the river's length that is flooded by dams.

The Tennessee River, for instance, is not a river. In North America the Appalachian Mountains run from the Canadian border some 2000 km southeast to the state of Georgia. The mountains rise to 2000 m above sea level and are largely covered in deciduous forest. To their east, short rivers from the Penobscot in the north to the Altamaha in the south drain the few hundred kilometers to the Atlantic. To their west (except in the north) the rivers drain into the Mississippi through the Ohio, its largest eastern tributary. In turn, the Ohio's largest tributary is the Tennessee. The Tennessee is 1000 km long and the fifth largest river in the United States in terms of flow. The Tennessee gets its name where two smaller rivers, the French Broad and the Holston, join near Knoxville in the state of Tennessee, just a few kilometers upstream of my house. But it is not a river, and my neighbors rarely call it one: It is Fort Loudon Lake. Downstream are Watts Bar, Chickamauga, Nickjack, Guntersville, Wheeler, Wilson, Pickwick, and Kentucky Lakes, named after the dams that form them. Almost the entire river is fragmented by these lakes.[15]

Dynesius and Nilsson show that almost all the major rivers are dammed, fragmented, and channeled by dykes along their banks to tame their flows.

To find a big wild river, one that isn't fragmented by dams, most of us have to travel. Europeans need only go to the Torneälven, the river that with one of its tributaries forms the border between Finland and Sweden. A convenient road runs along the Finnish side for easy viewing. It starts in boreal forest along the Baltic, then goes northward across the Arctic Circle and upward into the tundra of the watershed that forms the border with Norway.

This is Europe's only big wild river. North America has a couple of dozen wild rivers that drain into the Arctic Ocean or the Bering Sea. The Yukon is the largest, with the Skeena, Copper, and Kuskokwim vying for second place. All but the Yukon are relatively short—less than 1000 km long.

The Copper runs into the Gulf of Alaska, just east of Cordova. Coming into the airport, the small plane crosses the river's delta, a wetland with hundreds of small channels and ponds dotted with geese and swans. Eastward from the airport the road goes nowhere, ending at the Childs glacier. Cordova has no road connection to the rest of Alaska. The glacier noisily calves huge chunks of ice into the river; the resulting waves strand salmon along the shoreline. For 60 km downstream the road runs along, then crosses the river.

It does not look like a river, but then I realize that I have not seen enough wild rivers to know what to expect. In June, most of the river's course is dry. There are many small channels, filaments that break and join, like braided hair. Nowhere is the water wide or deep. When the snow melts in late spring, the river must be much more impressive. It leaves behind the boulders and large piles of tree trunks and other woody debris strewn on its wide, sandy banks. Farther downstream, the fishers wait for the annual runs of sockeye and coho salmon that give them their livelihoods.

To see a long wild river is an adventure. To cheat, look downward from a Tokyo-London flight as it crosses first the Amur, then the Lena, the Yenesey, and the Ob. Of the rivers cataloged by Dynesius and Nilsson, only the Mississippi is of comparable size. All these Siberian rivers are fragmented by dams, but the main channels of the Amur and the Lena are not. These two rivers' adjacent basins drain a Europe-size area of China, eastern Russia, and Mongolia, as the Amur drains into the Pacific Ocean and the Lena flows into the Arctic. From the air, sunlight reflects from their wide, braided channels all the way to the distant horizon. They must be extraordinary places on the ground. Neither river is easy to visit, particularly the Amur; it forms the uneasy frontier between China and Russia. May I recommend the Trans-Siberian railway for those with plenty of time on their hands and who lack the sense of immortality necessary to fly the Russian airline, Aeroflot?

The point of this connoisseur's guide is that wild rivers are few and remote. Future dam builders might take note: Almost all flow into the Arctic Ocean. Over Europe, east to the Urals, and North America, south of Alaska, almost every major river has dams for at least half its length that affect water flow. Many, like the Tennessee, have their entire lengths dammed.

All rivers that have parts of their basins in deserts and grasslands are severely fragmented. These are the regions that most critically depend on irrigation.

Another way to measure human impact is to look at what fraction of the river's annual flow is held in reservoirs. Reservoirs hold water for irrigation, and we would expect them to be important in dry areas. The Colorado, which drains the dry desert in the southwestern United States, leads the pack, holding six times its annual flow. It scores high in two other statistics too. Some 64 percent of its annual flow is used for irrigation and 32 percent evaporates from the river. Only two other rivers are comparable: the Amu-Dar'ya and the Syr-Dar'ya of central Asia, from which irrigation takes 50 and 80 percent of their water, respectively.

The Colorado has lost almost all its water by the time it reaches its last 100 km, the stretch that flows—hardly the right word—through Mexico into the Gulf of California. Pity the poor fishers so close to the bright green lawns and swimming pools of southern California and so far from their leaders in Mexico City who might help their plight. The fish are gone. They needed the river's nutrients and the food chain it supported. Gone, too, is the water. Their fishing boats are stranded in the sand of the dry riverbed.

Life in the River

Rivers bring sediment and nutrients (such as phytoplankton) to the oceans. The most productive fisheries are inshore, many of them near river deltas. Taking the water (and its nutrients) from rivers harms near-shore fisheries. As an example, start at the southern end of the Pacific Coast of North America and work northward to the Bering Sea.

The Colorado no longer has any water. The Sacramento and San Joaquin enter the ocean in San Francisco Bay, but they contribute more than 60 percent of their water to the agriculture of California's Central Valley. The 40 percent that remains contains agricultural and industrial wastes, and sewage. The once-valuable commercial fisheries of salmon, striped bass, and Dungeness crab are now history. The Sacramento's winter run of chinook salmon declined from 120,000 fish in the 1960s to 400 in the 1990s.[16]

The Klamath and Columbia rivers are extensively dammed. Salmon are anadromous, meaning they are born in rivers and streams, grow in the ocean, and return upstream to spawn. Dam the river and the salmon will have a hard time of it. In the Columbia River over 90 percent of the young salmon, born from the parents who made it upstream, die on their way downstream.

Finally, we have the Fraser River, which reaches the ocean at Vancouver, in the Canadian province of British Columbia. Going northward, there is no larger river south of the Bering Sea. The Fraser's tributaries have dams, the main channel does not, and it has arguably the world's richest spawning ground for sockeye salmon. After a year, the small fish, "fingerlings," swim downstream. They spend 3 years off the coasts of British Columbia and the states of Alaska and Washington, then return to spawn.

In 1996, Canadian and U.S. fishers caught nearly 11 million sockeye salmon—worth $100 million. In 1997 talks to revise the Pacific Salmon Treaty of 1985 broke down. U.S. and Canadian officials could not agree on the limits. Canadian fishers accused U.S. fishers in Alaska and Washington of unfairly overfishing Canadian-bound salmon. The local fishing fleet blockaded an Alaskan-owned car and passenger ferry in the province's port of Prince Rupert. And the province's premier wrote to the U.S. president informing him of his constituents' "horror, anger, and disgust" over Alaska's fishing of sockeye salmon.[17]

The lesson from all this? Wild rivers are now rare, their resources have consequently become much more valuable, and the conflicts over them have become correspondingly bitter.

The fate of these North American rivers is not unique. Dams on the Danube have changed the Black Sea's phytoplankton, favoring blooms of poisonous kinds. Those on the Volga have harmed the Caspian's lucrative caviar trade. The Aswan Dam on the Nile harms Mediterranean fisheries and coastal lagoons, and increases shoreline erosion. The lapse of an international agreement on the Farakka Dam on the Ganges in India reduced the flow to downstream Bangladesh by three-quarters, turning once fertile farmland into deserts and allowing salt water to intrude farther inland. Dams on China's Yellow River reduced flow by half, and, in 1995, the river mouth was dry for four months.[18]

Two central Asian rivers, Amu-Dar'ya and the Syr-Dar'ya, provide the most dramatic stories of humanity's impact. The Republics of Kyrgyz, Tajikistan, Turkmenistan, and Uzbekistan are among the 16 countries worldwide that irrigate more than half of their croplands. They share the two rivers with their larger northern neighbor, Kazakhstan. Both drain into the Aral Sea. In the Aral basin, over 90 percent of the harvest comes from irrigated lands.

Both rivers begin on the northern flanks of the Himalayas. The Amu-Dar'ya drains the Hindu Kush of Afghanistan in the southwest and the Pamirs of Tajikistan in the northeast, then runs approximately along the border between Turkmenistan and Uzbekistan. The Syr-Dar'ya begins in the Tien Shan mountains farther east and runs northeastward through Kazakstan, a little north of its border with Uzbekistan.

The Aral Sea—really a huge inland lake—was once the world's fourth largest freshwater lake, smaller only than the Caspian Sea, Lake Superior, and Lake Victoria. The destruction of this lake, an area about the size of Ireland, was planned from the outset. Soviet planners decided that the rivers' water would be worth more if diverted to irrigate agriculture. They ignored scientists who warned of dire consequences. The lake level began to fall in 1960. By 1987 the lake level had fallen by 13 m, its volume by two-thirds, its area by 27,000 km^2—roughly the size of Belgium or the state of Maryland. The exposed lake bottom is covered in salt and the remaining lake's salinity rose threefold.[19]

These changes are reflected in dramatic images, ones I found easily by entering "Aral Sea" into the search engine I use to access the World Wide Web. With this baseline, I compared the northern part of the lake to a declassified black-and-white satellite photo from 1962.[20] The date surprised me. Most American rockets exploded on or after liftoff until the early 1960s; this grainy photo was a technological triumph for its day. (Its justification was no surprise: The Soviets were keeping secrets in Central Asia that the Americans wanted to know about.) In 1962 the lake was not obviously different from when Commander A. Butakoff of the Imperial Russian Navy had surveyed the lake in 1848 and 1849 and published his results with Britain's Royal Geographical Society.[21] The Web site had a crisp satellite color image from 1987 of the same area. By then the falling lake levels had cut the lake into two: Small Aral in the north and Large Aral in the south.

I couldn't find more recent satellite images on the Web, but the search quickly pointed to a geographer Daene McKinney, who studies the Aral Sea. Off went an email inquiry to her at the University of Texas. Her reply came the next day from the far side of the world: "I am in Central Asia—Tashkent—at the moment with Philip Micklin," she wrote. "The lake has not yet disappeared, but it is much smaller than it was in 1988. The east and west portions of the Large Aral are now almost split."

It was Philip Micklin who in *Science* in 1988 predicted that the lake would shrink another 18,000 km^2 in the next decade, splitting Large Aral in two.[22] From the field, Daene had told me that Micklin's prediction was correct; the lake was on its way to becoming "a briny remnant in the next century."

Micklin appropriately subtitled his paper "A Water Management Disaster in the Soviet Union." Even by 1987 winds whipped up blowing salt and dust from the exposed lake bed. Major storms occurred almost once a month. Blowing dust and the declining quality of drinking water greatly increased the incidence of disease, particularly among children.

As the Aral shrank the fish disappeared and commercial fishery ceased, taking with it 60,000 jobs. Even the climate changed. Some 100 km south of the lake, humidity was lower, May temperatures were higher, and the growing season was too short to grow cotton. Cotton was the region's most valuable crop and the reason for diverting the water from the Aral in the first place.

Water That Returns to the Rivers

Not all water evaporates before it reaches its original destination. The global total of 1 trillion tons of borrowed water returns to the rivers, bringing with it the salts that remain after the evaporation of the nearly 2 trillion tons of used water. This returning water has now passed through the soil twice: once as runoff and once as used runoff. It can be very salty. The best irrigation water from large rivers may contain only a 0.25 to 0.50 g of salts per kilogram of water. The worst irrigation water may have 10 times that amount. At times during the summer the Red River of Texas and Oklahoma is more saline than seawater. In California's Central Valley, the used runoff has selenium salts with concentrations poisonous enough to kill wildlife in a wetland reserve created to hold the dumped water.

The used runoff also has fertilizers (nitrogen and phosphorus) and pesticides in it, for it has passed through soils of intensely cultivated crops. The natural levels of nitrogen and phosphorus in rivers vary a great deal, depending on the land through which they flow. Nonetheless, in wild rivers flowing through areas with little or no agriculture, a few parts of phosphorus to a million parts of water (ppm) is a typical value. When river basins are principally croplands, the values can be 10 times greater or even higher. Many of Europe's rivers exceed 10 ppm, some more than 40. North America's Mississippi is about 20, Asia's Indus over 10, and South America's Paraná nearly 10. The increases in the levels of nitrogen are broadly comparable.

Some of this nutrient-rich water ends up in freshwater wetlands. Wetlands are among the most productive terrestrial systems; they are not short of water (at least for some of the year); and the natural inflow of water brings with it nutrients that fertilize the plants. Nonetheless, there can be too much of a good thing.

In the Everglades of south Florida, the nitrogen and phosphorus from the agricultural area upstream are unwelcome because the natural water contains only low concentrations of them. The Everglades is one of the largest wetlands in the world. Water starts its journey in north central Florida, near Disney World, and flows south through the Kissimmee River into Lake Okeechobee. The river is now straightened and channeled—there is something about a winding river that engineers despise—and the lake is surrounded by levees to control its boundaries. The purpose is to provide water for Miami and the agricultural area to the south of the lake. South of the agriculture is what remains of the Everglades ecosystem. Across its northern border with agriculture and along its canals, dense stands of bulrushes thrive on the agricultural pollutants, replace the natural ecosystem, and destroy wildlife that depends on it.

Most of the water that flows off the land runs into the ocean, rather than into inland seas like the Aral, or through freshwater wetlands like the Everglades. Where it enters the sea, the rivers' nutrient-rich sediments create coastal wetlands, or salt marshes, above the low watermark and fertile estuarine environments below it. Estuarine ecosystems are among the most productive on the planet. They are also among the most polluted.

The Mississippi River drains more than 3 million km^2, delivering its water, sediments, and nutrients to the Gulf of Mexico. That basin contains almost 2 million km^2 of some of the most productive agricultural land in the world.

Nancy Rabelais, with the Louisiana Marine Universities Consortium is offshore in her boat, scuba tanks and wet suit at the ready to find out what happens as this water enters the Gulf. The freshwater floats over the Gulf's saltier, denser seawater. This layering intensifies in the summer and prevents any oxygen in the upper layer from reaching the bottom. The upper layer is rich in nitrogen and phosphorus from the widespread application of fertilizers on croplands that drain into the river, which is why these croplands are extremely productive. These nutrients stimulate the growth of phytoplankton in the upper layer. As they die they sink and decompose, a natural process that depletes what little oxygen was contained in the deeper layer.

The phenomenon is technically known as *hypoxia*, although it's more familiarly known as the "dead zone." When the oxygen concentrations drop, the fishing trawlers fail to capture any shrimp and bottom-dwelling fish. The low oxygen concentrations drive away the animals that can swim;

bottom-dwellers, such as shrimp, crabs, snails, clams, starfish, and worms that cannot escape, suffocate and die.

Hypoxia is not restricted to the Mississippi River. The rivers that drain into Europe's Black Baltic and Adriatic Seas and into eastern North America's Chesapeake Bay and Long Island Sound do similar harm.

Water Is Not Everywhere

Water, more than any other commodity, illustrates the maxim that "to him that hath is given more." The wet places of the world don't need the runoff because they are wet. The dry places of the world need the runoff to irrigate the land, but they don't get much.

So what do we do to the planet's rivers when we take the runoff that we need? There is no simple way to answer that. We use about one-sixth of the total runoff, but we do much more that does not appear in such a simple statistic. Across the northern third of the planet, the most densely populated and most industrialized third, we convert almost all the large rivers with dams and channels. Some large rivers are dammed along their entire length. The large rivers running through the dry lands are almost completely modified. We take all the water from some rivers. These physical modifications are not the whole story. We pollute rivers with nitrogen, phosphorus, and a vast array of by-products from our human activities. To a rough approximation, we do this in proportion to our use of the land in their watersheds.

Rivers connect the land to the ocean. Rivers also connect our impacts on the land to our impacts in the oceans. And it is those impacts that we must now uncover.

Part Two

BLUE OCEAN, GREEN SEA

From the top of Kaua'i, the northern Hawaiian island, one September morning, Paul Ehrlich and I looked out from the dark green forest of the Alaka'i swamp 1000 m down the pali to the ocean. "What percentage of the ocean's productivity does humanity use?" was the question on our minds. It is the same question as for land applied across the remaining two-thirds of the planet. This is a question of more than academic interest. Are the oceans the last refuge from human impact? Some hope that the percentage will be small, and in that answer lies the solution to humanity's population problems.

Five years later I am in Berlin for a conference and am about to meet people with such hopes. Dr. Silke Bernhard is our host and a legend in the scientific world for the precision and punctuality she instills in working, eating, and concert-going at the Dahlem conferences she organizes. Her perfected recipe is to forbid the formation of cliques of like-minded individuals that typify most conferences. She places me in a group with some medical doctors, a group I had rarely encountered at scientific meetings.

Together we were debating an editorial in *The Lancet*, the prestigious British medical journal. It dealt with the consequences of sterile oral rehydration packets. These are sugars and salts that children need if they are to survive dysentery and related diseases. Worldwide, most children do not get the help they need, and those diseases are leading causes of human death. The gist of the editorial was "What happens when these children grow up?" "Do they then not starve to death in even greater numbers? Is allowing the children to die the better choice?"[1]

The doctors felt they had an easy answer to this question. It is a question no one can answer, least of all people professionally committed to saving lives. "Use more of the planet to grow food!" they recommended. I replied by summarizing the contents of the first section of this book. The doctors were unaware that we had already co-opted a great deal of the planet's annual plant growth or that so little useful land remained. They thought that expanding the area of agricultural land would be a cinch.

"Then the solution to our dilemma is the oceans," one doctor ventured. "Aren't there still plenty of fish in the sea?" I could not answer that question easily, so I looked to the other ecologist in the group, Daniel Pauly, a fisheries ecologist then working in the Philippines.

His answer, then in embryonic form, was elaborated some years later in the pages of *Nature*.[2] Working with Danish marine ecologist Willy Christensen, Pauly produced two calculations for the oceans to parallel those of the Stanford group for the land.

Their first calculation is of how much green stuff the oceans produce. Unlike the land, the oceans' plants are phytoplankton. The method of scissors and undergraduates will not work for them, except along the ocean fringe where seaweed grows in shallow water. The world's oceans are bigger, deeper, and far more difficult to explore than is the land.

How do we know where to take the measurements? The ocean, like the land, has its barren deserts—the blue waters far from land—and its productive oases, the green seas. I needed to talk to ecologists who knew about the two-thirds of the planet that is completely unknown to me. To find them I had to enter their domain: the ocean and the research ships on which they visit it. That adventure is described in Chapter 8.

Pauly and Christensen's second calculation was how much of the oceans' green stuff we eat. Unlike the land, most of what we take from the oceans is not plant life, nor even animals that eat plants, but the ecological equivalent of lions and tigers: animals that eat the animals that eat the plants. They calculated how much of the oceans' green stuff feeds the small animals that feed the fish that feed us. Those calculations constitute Chapter 9.

Whether on land or in the sea, plants are green. The blue oceans produce very little green stuff and what they do produce is scattered over vast areas; we can use little of it. Green seas are mostly coastal and shallow. Pauly and Christensen's numbers show that a third of the oceans' production goes to support our fisheries—a figure broadly similar to the two-fifths of the land's production. That is a large percentage and certainly does not offer unlimited hope for oceans to be a new source of food for humanity. Nor is it one that suggests that the oceans are unaffected by our activities.

The evidence is to the contrary. Many fisheries have collapsed. Whales were overhunted in the last century. (Those who exploit the sea are rarely taxonomically correct; our harvests of whales, crabs, and scallops are all called fisheries.) The former New England whaling towns of Mystic and Gloucester are now tourist attractions. California's Cannery Row on Monterey Bay, known from the book and the film, and now for its trendy restaurants, hasn't processed sardines in a long time. Cod have gone from Newfoundland; its fishers live on welfare checks. And prepare yourself for shocking news. Caviar isn't always what it used to be; even from reputable sources, some of it is fake. Roe from other kinds of sturgeon, including the endangered kaluga from the Amur River, is passed off as the most valuable beluga, sevruga, and osetra caviars. The sturgeon that produce beluga caviar have been overfished, and there are even fewer kaluga, according to Ellen Pikitch, Director of Marine Programs of the Wildlife Conservation Society. She and others have called for a moratorium until wild sturgeon can recover.[3] Whether it is the laborer's (Atlantic) cod and chips or the socialite's (freshwater) sturgeon's caviar and blini, fisheries are in trouble.

Why are we so inept that we drive such natural and valuable resources to near-extinction? Why don't we manage fisheries in a sustainable way? Chapter 10 offers some explanations.

We reach the same conclusions for the oceans as for the land. Humanity does not use all of the oceans' production, but what we do use is already enough to damage the oceans, seriously and perhaps permanently.

8

On the Hero's Platform

It was July 1996, and Kendra Daly was not at her best. It wasn't that it had been 9 hours since she had eaten a hot meal, because the food was good and peanut and jelly sandwiches always filled the gap. She was generally comfortable, even if the top slot in the three bunks was too close to the ceiling and she would sometimes bang her head. Nor was it the hour, 1 AM, though lack of sleep certainly did not help. Four hours snatched whenever possible had been her schedule for the last 10 days and would be for another 30. Rather, Kendra was missing her drug of choice.

She could work without it; she did not particularly enjoy its effects. It made her mouth dry, her eyes unfocused, and her mind so fuzzy that she had to take special care not to make errors. She understood intellectually why it was no longer available, even with a doctor's prescription. But as she made her way outside into the blowing snow, toward the hero's platform, grabbing every handhold as the ship rolled, pitched, and yawed, she missed the relief the scopolamine patch gave her.

It was her hour at "the wire," a 1-by-2-m "platform" that was actually a grating lowered over the side of the ship by chains and that enabled access to a cable along which were strung the precious PVC Niskin bottles containing the samples on which the rest of Kendra's year's work would depend. The derivation of the platform's nickname is obvious. Working on the wire is a harrowing experience, even when wearing a hard hat and life-jacket and strapped into a safety harness. Kendra is measuring plant production in the twilight of an austral winter, 67°S in the sea ice off the coast of Antarctica. She has been collecting these data for over a decade, in summer and even in winter, when in the few hours of daylight she would watch the seals, whales, and penguins in the leads between the ice.

This chapter is about how Kendra and others take measurements, decide where to take them, and assemble them into a comprehensive report that creates an estimate of the ocean's total plant production. The simple answer it presents is parallel to the one in Chapter 1: How much biological interest do the oceans produce each year? It is the first of the two numbers that Daniel Pauly and Willy Christensen calculated.

The Oceans' Productivity

Measuring plant production in the oceans is a specialized task. In the oceans, tiny phytoplankton take up bicarbonate from the water, not carbon dioxide from the air, as trees do. The first task is to decide where to collect the phytoplankton. There are a lot of choices since oceans cover two-thirds of the planet. Getting to all but the thin line near land's edge requires a special boat, a skilled crew, nerve, and lots of research funds. Trips like Kendra's take years to plan and involve dozens of scientists. At every carefully selected station, each researcher wants a turn at the wire to load and unload sampling containers.

The Niskin bottles are lowered overboard on the wire and then they open and close at predetermined depths to take in phytoplankton, bacteria, and small animals. As depth increases, light decreases, and so does plant growth. To estimate the phytoplankton's total production, the productions at different depths must be summed from the ocean's surface down to where it becomes too dark.

Kendra's work begins when she comes off the hero's platform. Into each sample of seawater she injects a small quantity of bicarbonate. In this particular bicarbonate the normal carbon atoms (carbon 12) have been replaced with carbon 14, a radioactive isotope. As the phytoplankton grow, they absorb this carbon into their tissues. Kendra puts her samples into tubes in a clear Plexiglas box on deck, one custom designed to hold them steady as the ship pitches and heaves. She covers the samples with different screens. For samples taken at the sea's surface, the screens are nearly transparent, as is the water. For samples from farther down she uses increasingly opaque screens to imitate the reduced light in deeper water.

Twenty-four hours later the samples are ready. She pumps each sample through a fine filter that traps the bodies of the phytoplankton. Finally, a scintillation counter measures the radioactivity that each filter has trapped. The result is the amount of carbon that the phytoplankton used for their growth at that depth, on that day, at one point in the vastness of the world's oceans.

This process is repeated over and over—another day on the hero's platform, another point on the map. Do this every day from pole to pole, January to December, in every part of every ocean and you will find the answer to this chapter's single question: How much green stuff do the oceans produce each year? As on land, it is an impossible task. Just as on land, different parts of the oceans differ greatly in their plant productivity. Practical considerations mean that ecologists have to sample representative parts of the ocean, the equivalent of productive forests and barren deserts.

Going to Sea

In the early stages of writing this book I knew I needed to write about the oceans, provide a quick tour, and put two-thirds of the planet into perspective. But I simply had no idea what the oceans and the seas were like. I'd never been more than a few hours away from land. For years there was a large gap in the book while the rest of the text took shape.

Yet here I am tapping away at my laptop as the National Oceanic and Atmospheric Administration's (NOAA) research ship, the *David Starr Jordan*, rides on 3-m swells several hundred kilometers off the west coast of Mexico. How did I get here? The prosaic answer is that I flew to Panama, boarded the ship, and have been traveling north and west for 10 days with another 7 to go before we dock in San Diego, California. Another answer is that Lisa Ballance and her husband Bob Pitman, both marine ecologists at the Southwest Fisheries Science Center in San Diego, asked me to join the research cruise they were leading.

Ecologists are a hospitable lot, often inviting each other to their field sites. Company is welcome. It gives them a chance to talk science long into the night, with only the modest cost of finding a place to stretch out a sleeping bag or sling a hammock. So when at a scientific conference Lisa had offered such a chance it didn't seem that unusual an offer. Only when she continued with, "I'd love to show you 9°N 90°W," did the novelty sink in. I realized that was a long way off the coast of Costa Rica.

Lisa is talkative, enthusiastic, and looks quintessentially Californian with her long hair threaded through the back of her turquoise baseball cap, emblazoned with our ship's name. Bob is bearded, quieter, more deliberative, and looks as though he has spent a lifetime thinking about the oceans. His red baseball cap, bearing the name of another NOAA research ship, and his collection of T-shirts from different cruises, provide quiet testimony to the fact that he has. Together, they have carefully surveyed a huge swath of the planet's surface, strips of ocean thousands of kilometers long and a few kilometers wide, from pack ice to pack ice, and from land to the seven seas' centers. All this has been done from the flying bridges of research ships, with binoculars in hand. One afternoon we calculated that Bob had scrutinized 3 million km^2 of oceans. He modestly declined any honor, naming others who had seen even more of them.

In their water world I am a visitor wanting to see it all, like some American tourist doing Europe in seven days, asking for the highlights, camera at the ready. I need them to explain what I'm seeing, to explain the ocean equivalent of a desert and a rainforest. And I need them to couch their explanation in terms of life forms that are familiar to me—birds, whales, dolphins, and the larger fish.

The Composition of the Sea

"The first major division lies at the shelf break: Waters over the shelf differ markedly from waters seaward of it," Lisa explains. "The shelf break is the geological edge of the continent, sometimes close to shore, sometimes hundreds of kilometers from it. Waters landward of the shelf are shallow, less than 200 m deep, green and turbid. At the shelf's edge the sea floor drops quickly to thousands of meters below the surface, where the water is blue and clear."

In front of us, we see a school of dolphins and Bob and Lisa break off our conversation to guide their team of six marine mammal specialists. Each watcher is studying the mammals, intently counting them, checking which kind they are, studiously taking notes for later comparison. I grab my camera and marvel at how clear the water is, and how far down I can see the dolphins when they come over to ride the ship's bow waves. They provide vivid proof of the point about the water's color and clarity.

Plants are green, so where the land's surface is white or yellow we know that it is too cold or too dry for plants to grow well. In the oceans, the phytoplankton are both wet and warm enough. What limits their growth are light and food. The food is the supply of nutrients that builds the plant's tissue. Carbon from the bicarbonate that forms when carbon dioxide dissolves in seawater is usually abundant. Nitrogen, phosphorus, and in some places iron and silica are so scarce that little can grow. The oceans are mostly blue, which indicates that these special nutrients are in short supply. On continental shelves, the nutrients that run off the land can make the water cloudy; these nutrients feed the green phytoplankton, which also reduce the water's clarity.

"The difference in production is obvious from the birds and mammals we see above the ocean or in its surface waters," Bob continues. Above blue waters, hours pass with no bird in sight. This provides a clue as to what is *not* going on in the waters below. Indeed, that is why we are having this conversation, when there is little to distract us. In the last three days, we have seen only a scattering of birds, lone storm petrels, small parties of phalaropes, and an occasional jaeger looking for weaker birds to rob of their food. I comment that we may not have seen more than a few birds per hour. "In the central Pacific, we've gone days and only seen a couple of storm petrels per day," Bob tells me. By contrast, shelf waters team with life.

I had visited shelf waters before, taking coastal ferries such as the one from Kodiak Island to Seward in Alaska. It was an overnight trip in June, and the flat-calm sea was an enormous relief since this can be a rough ride. The sun lingered above the northern horizon almost all night, spreading its orange glow across the water and the mountain glaciers until they fell below the horizon. In every direction, I could see whales spouting. Black-and-white Dahl's porpoises swam by and the sea was crawling with fulmars, short-tailed shearwaters, and a wealth of other seabirds all new to me.

Other coastal waters were familiar because I had snorkeled in them. Coral reefs are increasingly familiar to everyone, as ecotour operators take customers to reefs off Florida or Australia to marvel at the bewildering variety of fish of every color and every color combination.

"On those boat trips you've taken to watch seabirds or whales from land, you're still seeing those typical of the continental shelves—shearwaters, gray whales, seals," Bob explains, and in doing so gently points out my terrestrial bias. "What most people think of as seabirds—gulls, terns, cormorants, brown pelicans—are birds of the coastal inner shelf, rarely straying from the sight of land. Only when you get off the shelf do you pick up the true oceanic life."

The fact that I have seen lots of such coastal birds and mammals impresses Bob and Lisa about as much as someone telling me about his forest experiences in New York's Central Park. Ocean life is special; you have to travel far from shore to find it. Even when their total populations are large, ocean birds have to fly great distances to find their widely scattered food. Over the vast extent of the oceans, these birds may be sparsely distributed. "That explains why we haven't seen the gadfly petrels I was hoping to see!" I complain.

"Ocean birds are also very patchy," Lisa says encouragingly, meaning when we see one, we may see several. The birds and mammals we have seen have been together in patches. Birds are often our first clue, milling around over the water's surface. Beneath the birds are schools of dolphins and tuna.

These predators drive the smaller fish to the surface, incidentally aiding the seabirds. Seabird flocks and jumping dolphins are how fishermen find the tuna in the ocean's vastness.

"The next major division is by latitude," Lisa goes on with the lesson, "high latitudes and the tropics."

I draw a box on my sheet of paper, labeling the two columns "high latitudes" and "tropics" and the two rows "shelves" and "off the shelves." Each one of the four boxes has its own characteristic life. High latitude shelves have gulls, terns, cormorants, alcids (birds such as puffins and murres), beluga and gray whales, fur and harbor seals. Tropical shelves have different kinds of terns, brown and blue-footed boobies, and some distinct kinds of coastal dolphins. Off the continental shelves at high latitudes are albatrosses, Dahl's porpoises, lone male sperm whales. And off the shelves in the tropics are red-footed and masked boobies, tropic birds, plus the still-elusive gadfly petrels, and, in the poorer waters to our west, sooty and white terns.

Our conversation ends at sunset, a predictably glorious one, with the watchers discussing the likelihood of a green flash. I thought the idea of the sun flashing green as it set was a myth, perhaps created by those who imbibed too many margaritas prior to the event. "Not so!" say my companions, who have seen sunsets far at sea by the hundreds, on ships where alcohol is strictly forbidden.

Our journey north and west from Panama has not always been smooth. Days of calm seas have been rudely interrupted by 4-m swells, when I've used Kendra's drug of choice (available again), experienced its side-effects, slept them off half the day in my bunk, and banged my head on the ceiling a few times. But I'm getting used to tapping away at my computer as the ship heaves and pitches, and it's time for another lesson.

"There's a third part to the story," Lisa continues one morning. "There are small-scale, physical features that cause waters to be much more productive."

"Such as?" I ask.

"Islands are an obvious example, rising from the deep ocean floor, as do the Hawaiian Islands. They locally affect the currents, which swirl and eddy after they flow downstream of the island, and they concentrate the nutrients and the phytoplankton that depend on them.

"Then there are sea mounts, which can divert deep, nutrient-rich currents toward the surface. A *seamount* is really just another kind of island, but one that doesn't reach the surface. They too cause downstream wakes and eddies."

"What about Stellwagen, off Boston?" I ask.

"That's a bank," Lisa corrects me. "It's on an already shallow shelf, but, yes, it does divert water upward." The consequences of that enrichment are familiar: The Stellwagen Bank is an easy trip from Gloucester or Province-town on Cape Cod. Where once whalers set off to kill whales, their modern-day descendents set off with boatloads of tourists to marvel at the whales feeding there.

"Upwellings are another physical feature." It was the Costa Rica dome upwelling at 9°N 90°W that had constituted Lisa's initial invitation. Lisa draws a sketch.

"The large current flowing north along the coast of northern South America is the Humboldt Current. Once it deflects to the west, it becomes the South Equatorial Current. The Costa Rica dome forms from the current to its north—the Equatorial Countercurrent—which flows in the other direction, from west to east. When the Equatorial Countercurrent hits the coast of Central America, all that water must go somewhere. Some is deflected to the south, some to the north, and a small portion forms a coun-terclockwise gyre."

"That's opposite of the clockwise rotation of the major flows in the North Atlantic and North Pacific, right?" I ask.

"Yes," Lisa replies. "Moving water and air experience the Coriolis effect—the apparent deflection to the 'right' (clockwise in the Northern Hemisphere) or the 'left' (counterclockwise in the Southern) of their path as a result of the Earth's rotation.

"As a consequence of the counterclockwise gyre, there is what is called Ekman transport: Water flows outward from this circulation rather than toward its center, as in the gyres of the central North Atlantic and Pacific."

"So that must pull water up from the depths, water richer in nutrients than water that has traveled the surface easterly across the Pacific," I ven-ture.

"Exactly!"

The Costa Rica dome had been a huge disappointment a week earlier. Its green waters were the reputed home of blue whales, other mammals, and birds I wanted to see. I had watched the ship's Global Positioning System (GPS) read out our position, closing in on the expected center of the dome, with no whales, no green water, and no bird flocks anywhere in sight. Nonetheless, now that we had been in truly blue waters for a week, I ap-preciated how many more birds and dolphins were in the vicinity. The experience showed that ocean currents are fickle and variable, a further complication in the patterns of ocean production.

"We'll see another example before we get home—the California Current." Lisa drew another sketch. "It flows southward along the coast of California to the southern tip of Baja California. The Ekman transport of water away from the coast, aided by seasonal winds from north to south, also pulls deeper, nutrient-rich waters to the surface."

"Blue whales?" I asked, hopefully.

"A definite possibility!"

"So what other features are there?"

"The edge of the ice. That's Bob's area. He has been to Antarctica several times."

Another blue water day, few birds, and so an opportunity for Bob to tell more of his story. I sit facing aft and into the sun to shade my computer screen. Reggae—Bob Marley—is blaring over the speaker to entertain the watchers who aren't finding any dolphins. It is a glorious day; the sea is a deep blue with whitecaps and swells that my newfound sea legs allow me to ignore.

Bob's trips had been on the Japanese ship, the *Shonan Maru*, which he joined at Wellington, New Zealand. She had been built as a whaler, was staffed by whalers, but took a small complement of international scientists with her, this time on a trip to Valparaiso, Chile. The ship is owned by a whaling company and funded by the Japanese government as part of its support of the whaling industry. To offset expected demonstrations by Greenpeace, a large, hastily constructed banner was draped over her side, proclaiming (truthfully) the mission "Research." Other scientific whalers would kill whales, but not this one. Their job was to count whales to gain a sense of how many there were.

"What was the boat like?" I ask, my total experience being my 10 days on the *David Starr Jordan*.

"Faster, capable of doing 15 knots to our 10, private rooms, well stocked with beer and Suntori whisky." (Our ship was strictly free of such temptations.) I guessed the hot tub correctly and the fact that the crew would smoke continuously, as do all my Japanese colleagues.

"About 10 in the morning, with the wind blowing off the ice and the wind chill −40 or worse, everyone was dreaming of the hot tub. The smoking was a problem. The only way to avoid it was to go outside in the cold or to isolate yourself in your room.

"It took us 6 days to get there. Through the roaring forties and the furious fifties. Horrendous weather, 50-knot winds (90 km/h), and 10-m swells. And when we got there, we had a week of fog during which time we couldn't see anything. Two-month trips can have 50 percent downtime when it is impossible to collect data. When it clears, there is sea ice. It doesn't always form a neat boundary, so the boat sometimes has to skirt small patches of it. Life is concentrated along this edge.

"Some areas were really rich, and we sighted 400 minke whales some days. Twenty were typical. Penguins and seals would haul onto the ice to avoid their predators, the killer whales."

The number of killer whales surprised Bob; he had never before seen pods of 60 to 70 of them. The minke whales, another favored prey, could only play a deadly game of hide-and-seek.

Krill, cocktail-shrimp-size crustaceans, were the minkes' food. Wine-colored patches of them, 10 to 15 m across, dotted the ocean. They, in turn, fed on the algae that grow on the undersides of the ice.

Glacier ice was prized by the crew, too. The dense ice was winched on board by the ton, kept in the freezer, some to be taken home as souvenirs, some to be broken into smaller pieces and dropped into the whisky. As it melted, the trapped air bubbles from centuries ago would give off a distinctive sound. Bob's souvenirs were his distinct memories:

"Two months without seeing another ship, an ocean free of flotsam."

Debris passed by every few minutes on the *Jordan*.

"The amazing colors of the first icebergs, an unforgettable deep blue, some of them kilometers across. And the haunting, diffuse light of the long summer days."

Light and Nutrients

While Bob and Lisa's descriptions of the oceans continued to fascinate me for the 3 weeks of the voyage, I needed a succinct summary.

"It all comes down to the two things that phytoplankton need: light and nutrients. Their distribution across the world's oceans essentially determines the biological "realms." The shelf is high in nutrients and the ocean is low. High latitudes have high nutrients and have seasonal light; low latitudes are low in nutrients and have constant light. And physical features either concentrate nutrients, or plankton, or both."

Bob and Lisa's classification did not set out to deliberately group the oceans into realms according to phytoplankton distribution and abundance. Their descriptions were of birds and mammals and the oceans' color—green or blue—descriptions not so different from what one can read in James Cook's journals written 2 centuries ago. Their classification focused on what we can see, and not on the microscopic plants, the zooplankton, or on the tiny animals that feed on them, or on the fish that exploit these animals and which we in turn exploit. All of these correlate, but it takes newer technologies to complete the picture.

Minute by minute, day and night, the pings from our ship's echo sounder measure where objects reflect the sound. A computer monitor, with a navy blue scale on either side in meters, displays the results. Moving across in real time is the density of objects that the sonar reflects. The image is a pointillist painting—lots of colored dots, yellow and green dots near the surface, then at 40 m down an intense band of red ones. Red represents the strong reflections, green the weaker ones, so the screen tells us that there's a lot of stuff at this depth. It is probably zooplankton that cannot move up and down the water column. Below that, at 80 m, is blackness, which means no reflections. Then at 220 m below the surface, a thin, yellow-green band, a broader red one (240 to 260 m), and beneath it, a yet broader green band down to 400 m, then nothing again.

All this changes spectacularly at sunset. On cue, the lower band—apparently lantern fish and squid—moves to the surface. The two separate sets of bands join for the night. Within half an hour of sunset, one dense band lies within 40 m of the surface; the rest is scattered down to 200 m.

So this is where life is in this ocean! Hundreds of kilometers from anywhere, and hundreds of meters below the surface, life comes to the surface only at night. During daylight, like Dracula, it flees downward again into the depths. The pattern repeats night after night.

The Big Picture

Ocean ecologists acquire their data the hard way, so knowledge about the oceans comes slowly. One day at a time, as dawn to dusk transects across the often near-lifeless surface, ecologists dip their nets into the depths to catch the small and elusive animals and plants.

By the late 1950s there were nearly 200 measurements of phytoplankton production worldwide, each taken at several depths. That is six measurements on average for an area the size of the United States or Europe—all to cover the huge range of seas and oceans that Bob and Lisa had described to me. It is as if I asked you where you would put your six measurements to capture the range of plant productions over the United States: its deserts, forests, mountains, and swamps. And, since you get only one day in which to take a measurement, which day do you pick? Nonetheless, the estimates were enough to show huge variation in production.

For instance, the Benguela upwelling off southern Africa produces the equivalent of 1.5 kg of biomass per square meter each year—as much as a productive ecosystem on land. A square meter in the central north Atlantic produces only one-fiftieth as much, which is the same as the driest deserts or coldest tundras on land.[1] The waters of continental shelves, the shallow water inshore of where the sea bottom drops quickly to the ocean depths, were intermediate. So, too, was production from rivers and lakes. Coastal waters, including coral reefs, were highly productive, too.

Any global estimate of productivity depends on how large the centers of production, such as the Benguela upwelling, are. How long do these areas remain productive? Are these values maintained day after day for 365 days a year? Or are they less in the shorter days of winter? Do the currents supply the nutrients at a constant rate? Or do they vary seasonally or from year to year like the ephemeral Costa Rica dome?

The problem is that the scale of the measurements is greatly at odds with our need to extrapolate to years and oceans. Can we extrapolate values taken from a few liters of water on one day to the whole ocean for years? Only more cruises, more surveys, more days on the hero's platform will tell.

By the late 1970s marine ecologists had expanded the data to include many thousands of data points worldwide. Each point still characterized a huge area, but patterns were beginning to emerge.

These samples report that, on average, each year a square kilometer of open ocean produces about 225 tons of plant biomass, an upwelling about 2000 tons, a nontropical shelf about 700 tons, a tropical shelf about 700 tons, and coastal water and reefs about 2000 tons.[2]

These different kinds of oceans vary greatly in the area that they cover. The open ocean covers 332 million km^2, upwellings less than 1 million km^2, shelves 27 million km^2, coastal waters and reefs about 2 million km^2, and rivers and lakes about 2 million km^2.

In the same way that we estimated the land, we can multiply these areas by average values of how much they produce. In billions of tons of biomass per year, these are the estimates: open oceans 89, upwellings 3, tropical shelves 6, nontropical shelves 12, coastal waters and reefs 4, and rivers and lakes 1. The total, 115 billion tons, was the number Pauly and Christensen used in their paper.

Do all oceanographers agree on these numbers? Has what oceanographers learned since the late 1970s significantly altered the numbers? Given the technical difficulties of finding life across and within the world's oceans, we expect that this hard-won information will constantly change as more details trickle in.

In 1995 Alan Longhust and his colleagues in Nova Scotia, Canada, computed the global production at between 105 and 120 billion tons per year.[3]

This is close to the same number, but the details are different. There are several reasons why.

On board ship in the early 1980s, oceanographers realized that extremely tiny phytoplankton called *nanoplankton* and even smaller *picoplankton* could pass through all but the finest filters.[4] These could be abundant in the open oceans and contribute significantly to its production. However, it was not only their small size that created a problem.

Like the tiny protozoan predators that eat them and the even smaller bacteria that decompose their dead bodies, nanoplankton live fast and die young. Within a 24-hour experiment, a molecule of carbon may have passed from the seawater to a nanoplankton, to a bacterium, and back into the water. The early experiments missed the contribution of these tiny phytoplankton.

Another technical advance found that the tiny phytoplankton could easily be poisoned by toxic trace metals in the Niskin bottles. Experiments had to employ ultra-clean methods to avoid poisoning the phytoplankton. All this new technology had to fit securely into the pre-equipped container on the ship's deck that serves as a temporary laboratory.

Using ultra-clean methods to eliminate this poisoning, scientists' estimates of productivity in the open oceans of the equatorial Pacific rose from about 225 tons per square kilometer each year to about 400 tons.[5] If such numbers were typical of other open ocean systems, they would make a large difference in the calculation of global production.[6]

The Fine Details

As the number of samples increased, the picture of plant production became more complicated. Across all oceans there are intertwined patterns of winds and weather, of turbulence, waves, eddies, upwellings, and currents that move and mix water, nutrients, and phytoplankton on scales from hours to decades and from a few meters to 10,000 km. In addition, a productive area might not always be there, as our experiences of the Costa Rica dome had proved.

The subtropical north Pacific had seemed one of the most constant places. The same kinds of zooplankton were sampled at the same abundance year after year and at stations hundreds of kilometers apart. Yet an El Niño altered the mixing of the surface waters and increased the production by 40 percent.[7]

The central Arctic Ocean was similarly thought to be barren. Yet studies like Kendra's under the ice and in the *polynya*—the open waters between the ice—showed that the Arctic Ocean is productive. Indeed, it is sometimes highly productive where the waters are nutrient rich. Studies of blooms of phytoplankton in the polynya of Arctic and Antarctic waters raised the estimates of global production.[8]

Knowing when and where to find plankton blooms is the trick. It means seeing what parts of the ocean are green. The Coastal Zone Color Scanner (CZCS) was flown into space aboard the Nimbus 7 satellite in 1978. Month after month until 1986 it measured the oceans' greenness, which indicated the amount of chlorophyll pigments in the surface waters.[9]

The CZCS images are stunning in their detail and are forever changing. These images show the broad patterns familiar to shipboard scientists. The coastal waters of Western Europe, from Norway south to Brittany and including the North Sea and the Baltic, are green. So, too, are the coastal waters of Iceland, Greenland, and North America from Cape Cod to Newfoundland. The waters off California (including Monterey Bay) are green, and so is a wide swath around Alaska across to Siberia. The Great Lakes of the United States and Canada are green. South of the route you would fly from, say, London to New York, the oceans are much less green. These more blue waters stretch along the U.S. East Coast, Europe's West Coast south of Brittany, and in the Mediterranean. In the Southern Hemisphere the coastal waters of South America and the Atlantic Coast of southern Africa are green. So, too, are the waters off Antarctica.

Within these broad patterns are even finer details. The Orinoco River delta is located along the coast of Venezuela, where the river flows into the southern Caribbean Sea just east of Trinidad and Tobago. This large river drains most of Venezuela and Colombia, and northern Brazil, bringing sediment and nutrients from as far away as the Andes. The Caribbean's currents carry Orinoco's freshwater to the northeast along the South American coast. The river's water is fresh, and thus less dense than seawater, and it is enriched with nutrients for phytoplankton growth. A satellite image shows a plume of productive waters flowing toward the island of Puerto Rico, to the west of the Leeward islands.

The image is static, taken as the satellite passed overhead. But its pattern of mixed green and blue waters suggests the dynamic, turbulent swirling and mixing of the cup of coffee that sits next to me. Gently stir the black coffee and then add a stream of white cream. The white wriggles, bulges, and then breaks into swirls, which slowly disperse into the now-brown coffee (see Map 11).

These rich details solve many of the problems that attend the severely limited number of shipboard samples. Rather than thousands of data points, satellites provide millions of them, showing peaks of biomass and how long they last. They tell oceanographers where to look and when. And they explain why a fixed spot in the middle of the Atlantic may be highly productive one year and not the next. In the second year, a ring of warm water may be over the spot, slowly spinning northeast and cooling as it goes.

Satellite images do not provide all the answers. As on the land, plant biomass "green" and plant production "new green" need not go hand in hand, but there is a good correspondence. Generally, the greener the ocean, the more productive it is. But this is not always true.

The satellites measure greenness in the surface waters, yet the phytoplankton go deep in the ocean. There may be more of them 100 m below the surface than at any other depth, and their maximum production may be well below the surface, too.

This and other complications mean that we cannot simply read off how productive the oceans are from how green they are. To utilize the detailed images we obtain from satellites we must check what each colored pixel represents in the ocean. A particular shade of green must mean the same distribution of chlorophyll in the ocean's water column.[10] The only way to check is the hard way, from the hero's platform.

Ocean Deserts, Ocean Oases

Even as the details emerge, the general conclusions remain.[11] The oceans' total plant production is about three-quarters of that on land yet it is produced over an area more than twice as large. So, even on average the oceans are not that productive. The average is rather misleading, because it is a mix of 90 percent blue oceans and 10 percent green seas.

For their area, the green seas—upwelling ecosystems and coastal waters—are highly productive. That explains the abundance of birds above their surface and the seals, whales, and fish below it. These areas, and the productive shallow waters of continental shelves, are geographically restricted indeed. They account for only about 30 million km^2. The blue oceans are not productive and contribute little despite their huge area.

As I wrote this one early December day, on the flying bridge, when no other ships had been sighted in a week, I began to understand how these numbers explain such divergent views of the oceans. Some, like the physicians at the Dahlem conference, wonder whether we could ever seriously impact the vast natural resources of the oceans. A century earlier a famous scientist of the Victorian age had asked that same question. His answer begins the next chapter.

Exclude the barren open oceans and what remain are productive oceans only about a quarter of the size of the productive areas on land. It is these rich areas that play the crucial role in supporting our fisheries. How big are the oceans? From a human-centered view, the answer is "very small": 30 million km^2, or about the size of Africa. That view is the key to understanding what fraction of the total production humanity consumes, what fraction it might possibly consume in the future, and how much impact we have on the oceans and their resources.

My voyage is about to end; the coast of Baja California is on the horizon, orange in the late afternoon sunshine. Kendra had told me how her grueling trip had ended. They had run northward along a line of 170°W and ended in the warm, calm oceans of the tropical South Pacific, off American Samoa. On leaving the boat, she found a warm, sandy beach with a Tiki bar that sold cold mixed drinks with fruit and little umbrellas. In a hotel with a ceiling far above her bed, she slept most of the next 5 days. Recalling her story reminded me how welcome a beer would be on the following evening.

As the sun set on our last afternoon, two blue whales appeared off the *Jordan*'s starboard bow. They remained close to the ship until just before the sun set, flashing dayglow green as it went.

LIST OF MAPS

Map 1. The world according to Mercator is not very helpful for global accounting. Areas on the map near the poles are relatively too large, while Africa and South America look small by comparison to Greenland.

Map 2. The Albers projection is an equal area one. Although the countries are distorted to different degrees, the same area on the map corresponds to the same area of the planet's surface, wherever it is placed. Africa now appears as the huge continent that it is. Greenland is still large, 2 million km^2, but it is at least a more modest size. This version of the projection does not show Antarctica, which adds another 13 million km^2 of ice-covered land. The black areas are "tundra" and "forested tundra" in the Olson vegetation classification.

Map 3. The world's croplands and cities.

Map 4. The world's temperate forests. The word temperate is the clue to our impact as human beings. Neither too hot nor too cold, we have long cleared these forests for our settlements and crops. Compare this map to Map 3 for croplands and notice the broad similarities. This map combines three of Olson's classes: "temperate broadleaf forests," "mixed deciduous and coniferous," and "woods and fields."

Map 5. Coniferous forests. This is a combination of Olson's "coniferous forests" and "taiga."

Map 6. "Tropical humid forests" in Olson's classification.

Map 7. The Amazon River (flowing west to east) shows up black and braided near its junction with the Madeira (coming in from the south) in this satellite image. Smaller rivers flow in from the north. White areas with dark shadows—mostly south of the Amazon—are clouds. The pale areas on the north bank are roads and fields on either side of them. Although these areas are still heavily forested, their convoluted shapes mean that large areas of forest are quite near the roads. The image is approximately 150 km wide. Manaus is just to the west but not on the image. (*From a Landsat 7 image; courtesy of the U.S. Geological Service*)

Map 8. Dry forests. This map combined "shrubs and trees, thorn forests" and "woods, fields and savannas" in Olson's classification.

Map 9. "Grasslands" and "shrublands" in Olson's classification.

Map 10. "Desert" and "semidesert" areas in Olson's classification.

Map 11. Ocean colors in the sea surrounding Tasmania indicate swirly cold water and warmer waters in a highly dynamic pattern. Different shades of gray are different water colors; the darker ones nearer the island are greener than those farther from it. The white areas are clouds. (*Reprinted by permission of NASA, GSFC Earth Sciences [GES] Data and Information Services Center [DISC]/Distributed Active Archive Center [DAAC] color figure modified to gray scale*)

Map 12. The number of species of passerine birds in the Americas. The lightest shade of gray (Arctic Canada, southern tip of South

America) has fewer than 32 species. Areas with between 32 and 64 species are to the south, in the Great Plains of the United States and the southern and dry desert areas of southern South America. The deciduous and mixed forests of eastern North America have from 65 to 127 species. From 128 to 255 species are found from Mexico southward across Central America and throughout most of Brazil. More than 256 species—the darkest shade—are found in the Amazon, along the Andes, and in the coastal forests of Brazil. In other words, most species are in the middle.

Map 13. Species with the smallest ranges (less than 100,000 km^2) are concentrated into special places called hot spots. These include islands in the Caribbean, the Galápagos, Central America, the Andes, and the Atlantic coastal forests of Brazil. Species with small ranges, however, can—and do—occur anywhere, including the very southern tip of South America and one, illustrated by the arrow, in the upper peninsula of Michigan. White: no endemics, palest gray: 1–5, medium gray: 6–10, darker gray: 11–20, black: 20+.

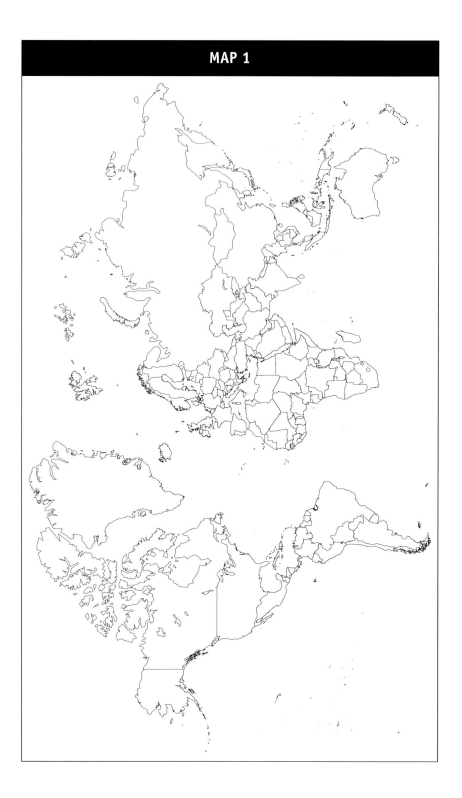

MAP 1

MAP 2

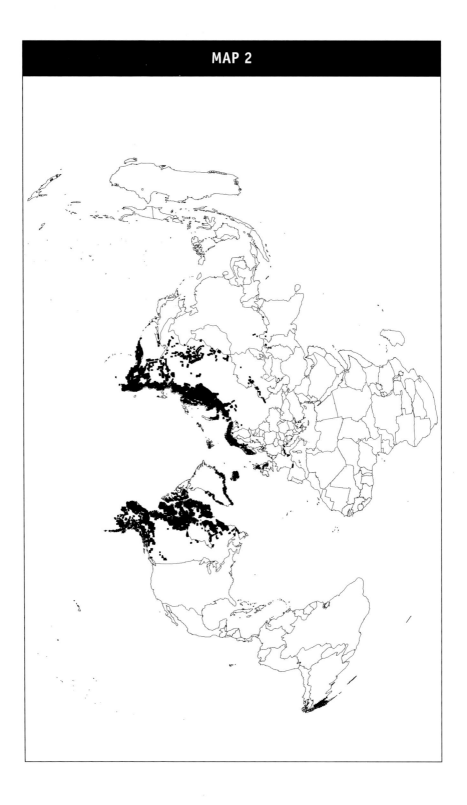

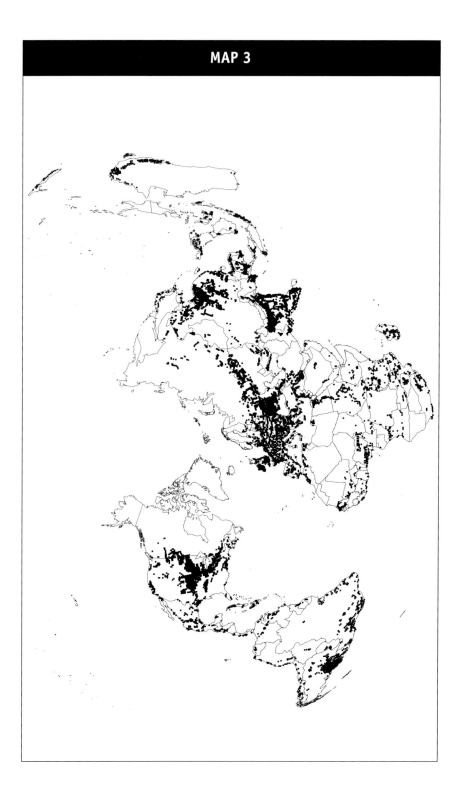

MAP 3

MAP 4

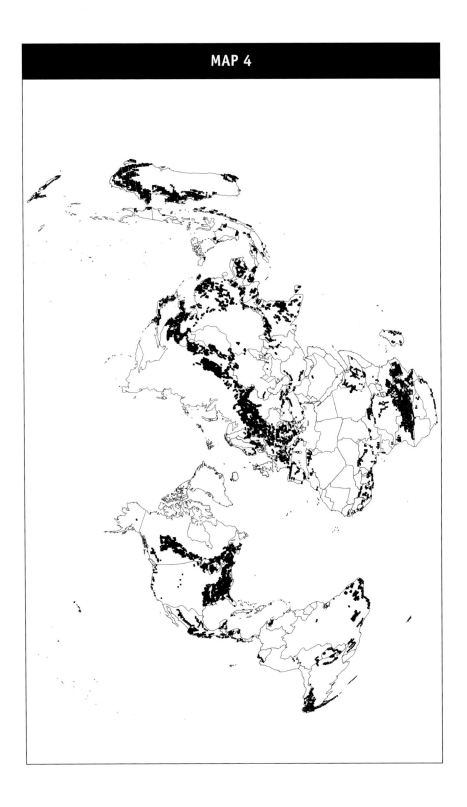

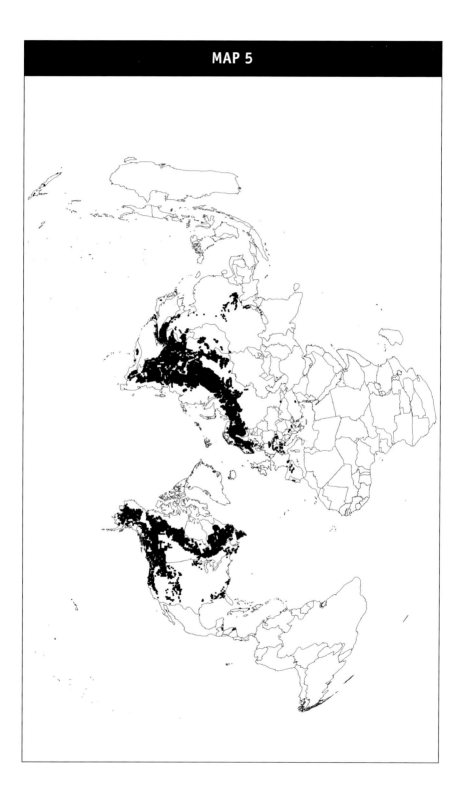

MAP 5

MAP 6

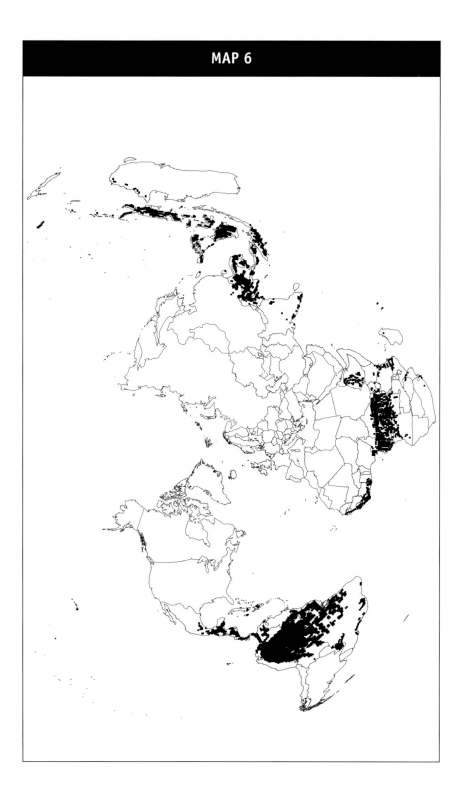

MAP 7

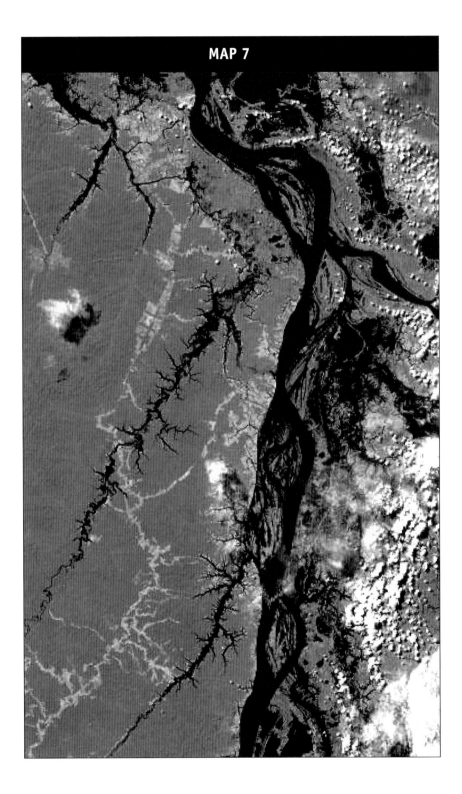

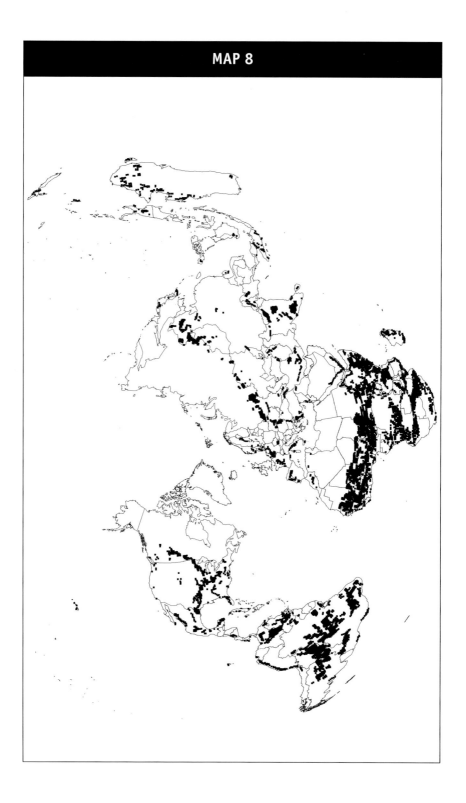

MAP 8

MAP 9

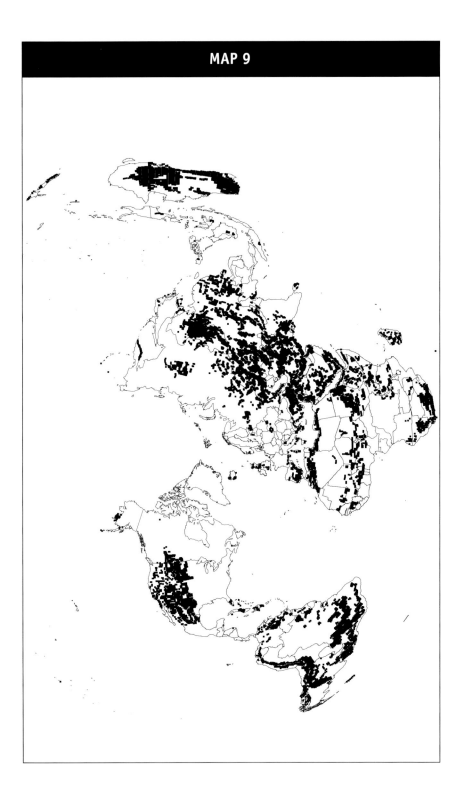

MAP 10

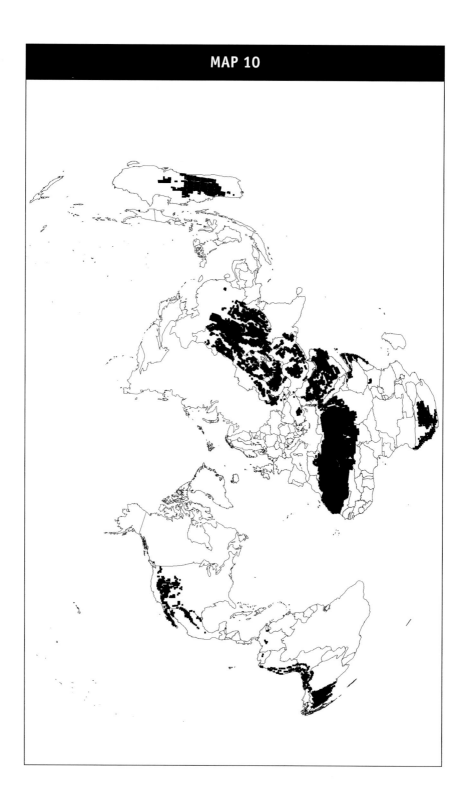

MAP 11

Southern Australia

Tasmania

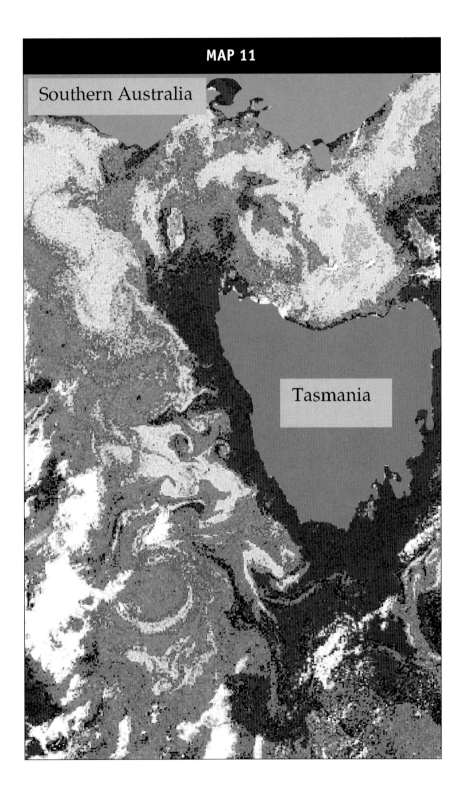

MAP 12

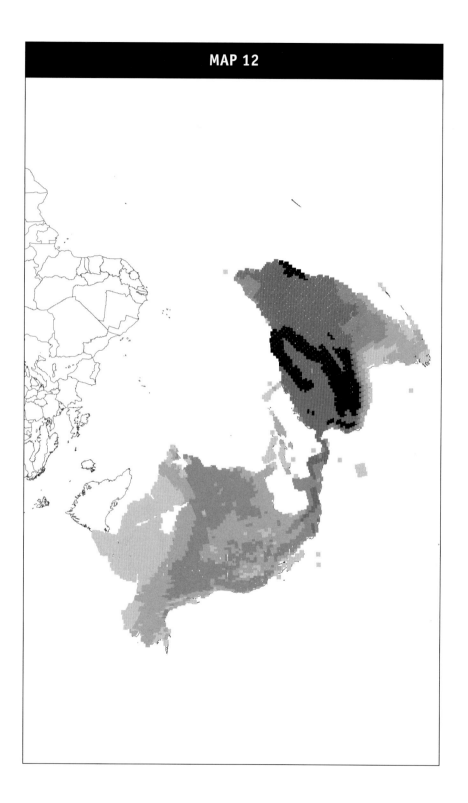

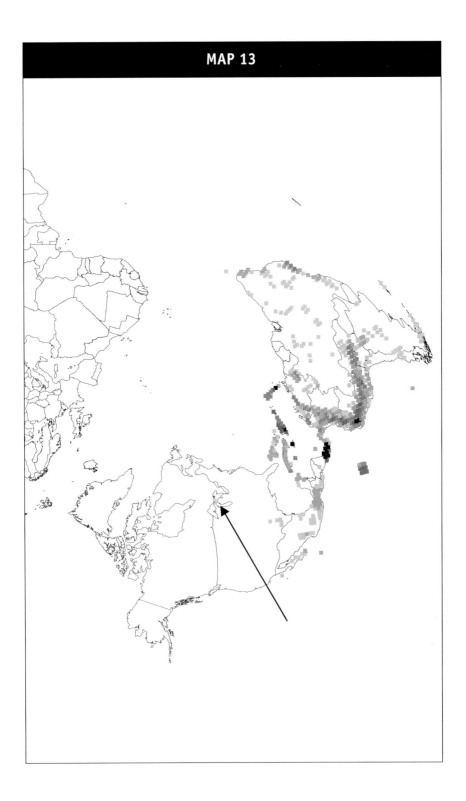

MAP 13

9

Lots of Good Fish in the Sea

London, June 18th, 1883: "Your Royal Highness, your excellencies, my lords, and gentlemen, it is doubtful whether any branch of industry can lay claim to greater antiquity than that of a fishery." So began Thomas Huxley's inaugural address to the Fisheries Exhibition.[1] Nearly 60 years old, his square jaw, mane of hair swept straight back from his forehead, and long white sideburns were appropriate for a man of his influence. He was "Darwin's bulldog," the public defender of evolution, while Darwin largely remained cloistered in Kent. Over a decade earlier, *Vanity Fair* had proclaimed of Thomas Huxley that "there is no popular teacher who has contributed more to the awakening of the intellect."[2]

To his audience, he posed the critical question: "whether fisheries are exhaustible: and if so, whether anything can be done to prevent their exhaustion." His question is this chapter's topic.

Huxley had devoted himself to these questions, he told his audience, and was glad to have the opportunity to submit his views to his audience's judgment. He continued, "I have no doubt whatever that some may be exhausted. Take the case of a salmon river. . . ." A salmon river *could* be destroyed and he pointed to pollution as a cause.

"And now arises the question: Does the same reasoning apply to the sea fisheries? . . . I believe . . . that the cod fishery, the herring fishery, the pilchard fishery, the mackerel fishery, and probably all the great sea fisheries, are inexhaustible; that is to say, that nothing we do seriously affects the number of fish."

Like the pollution of salmon rivers, much of what followed in his talk would sound familiar to us today. His units of production "hundred weights per acre" are dated, but suitably converted his estimates are familiar. His

145

description of oceans in the late 1800s, however, is strikingly unfamiliar. "At the great cod fishery of the Lofoten Islands [Norway], the fish form 'cod mountains,' vast shoals of densely packed fish, 120 to 180 ft (35 to 55 m) in vertical thickness. . . ." The weights on the fishers' lines would constantly knock against fish before reaching the bottom.

Huxley's predictions about that key advance in Victorian technology, refrigeration, were strikingly perceptive. In his boyhood, the salmon on the streets of London were from Newcastle and pickled. "My son, or at any rate my grandson . . . may be offered his choice between a fresh salmon from Ontario and another from Tasmania."

A visit to my fishmonger turned up more than two dozen items that would have delighted but perhaps not surprised Huxley. They included coastal oysters from Virginia and mussels from Maine, inner shelf crawfish, shrimp, grouper, and red snapper from the Gulf of Mexico, swordfish and albacore tuna from far out in the tropical oceans, and orange roughy from seamounts in the southern Pacific. On that visit I failed to find cod, haddock, herring, and many other fish that would have been familiar to Huxley. He would be puzzled if he went shopping today, for these are the fish that he thought would be inexhaustible.

Huxley's arguments, wrong though we know them to be, are interesting precisely because they are wrong. They tell us why arguments that the Earth's natural resources are inexhaustible are so quickly shattered by experience.

For 1993, the FAO listed the top 15 fisheries worldwide, in millions of tons of wet fish.[3] They were anchoveta 8.3, Alaska pollock 4.6, Chilean jack mackerel 3.4, silver carp 1.9, Japanese pilchard 1.8, capelin 1.7, South American pilchard 1.6, Atlantic herring 1.6, grass carp 1.5, skipjack tuna 1.5, chub mackerel 1.5, common carp 1.4, yesso scallop 1.2, yellowfin tuna 1.2, and Atlantic cod 1.1.

These catches are recorded as wet weight—the weight of the fish as it lands on the dock. One of the difficulties in global environmental accounting is that each group of scientists uses a different measure. Fishery statistics are always reported in wet weights, and I follow that tradition. When a comparison with phytoplankton production is needed, I convert the wet weights to dry weight to match the numbers I have presented in earlier chapters. (One has to divide these numbers by 4.1; and, no, I did not do this calculation by drying fish in my oven at home![4])

Fish landings vary from year to year. Like sports teams in a league, some fish consistently remain in the top 15 but change rankings, while others do well for a few years and then drop out of the rankings as others are added.

Are these numbers accurate? It is not easy to confirm them; we can only rely on what the fishers report most of the time. For some tropical coastal fisheries where thousands of fishers land fish in hundreds of villages along poorly developed coasts, the estimates are no more than inspired guesses. Sometimes there may be incentives to exaggerate the catch to make it seem more economically important than it really is. By contrast there are widespread inducements to underreport. Quotas often limit catches, with penalties when they are exceeded. Or there may be fees to pay for a catch in another country's territorial waters. And a fisher who catches fish of the wrong kind, or of the right kind but too small, of the wrong sex, or at the wrong stage in their life cycle, may not be inclined to admit it. Recorded numbers are likely to give a low estimate of our true impact on the oceans.

My cabin mate on the *David Starr Jordan*, Wayne Perryman, had spent 20 years as a NOAA Corps commander, most of it in the Bering Sea. His ship's mission was to work with the fishers to find out where and how they caught fish, and also to provide independent assessments of the state of the fishery. Such efforts provide cross-checks on the reported catches, something not all nations can afford to do, of course.

Whatever the uncertainties in the numbers and however they vary from year to year, the world's fisheries are always extraordinarily varied. Carp are freshwater fish farm-raised in ponds, mostly in China. The rest are marine fish. Anchoveta are from upwellings, tuna from the open oceans, and cod and pollock from shelf waters. Their habits are different, and we catch them using different methods.

These differences are at the core of the problem of estimating our impact on the oceans. On land, the calculations of impact were simple. We consume plants either directly—harvesting crops, trees, or burning forest, for example—or by having our domestic animals do the harvesting for us. In the oceans our use of plant production is almost entirely indirect; we eat fish that eat plants. It is also varied because we eat fish that eat zooplankton that eat plants. And we also eat the fish that eat the fish that eat zooplankton that eat plants!

Going Fishing

The largest fishery, anchoveta, comes from one of the oceans' most productive areas, the upwelling off Peru and Chile. Catches vary greatly from year to year. In 1970 the catch was 13 million tons, but it crashed to 2 million tons in 1973 and now has recovered.[5] These small anchovies swim in dense schools. As adults, they develop large mouths to sieve out the phytoplankton from the water. When younger, they feed more on the zooplankton. They are caught in purse seines, nets dragged behind a boat that scoop up the fish. The fish are dumped unceremoniously into the hold, without refrigeration.

"Very much a coastal fishery, with small ratty boats," was Bob Pitman's description. "The boats return to harbor within a few days." Bob had seen them coming into Iquique, Chile. "In such a climate, the fish quickly begin to rot. Their decomposition is so fast that the pile of fish is smoking. The ammonia fumes that come off are lethal. The workers who clean out the fish from the holds with a large suction tube have to wear hoods and oxygen tanks to do so."

Some of the fish are caught in the net by their gills, and these fish are left to rot. Those sucked from the hold are piped straight into factories where they are converted into fish meal fertilizer. Anchoveta are not worth much, less than $100 per ton. Only a tiny fraction of the catch ends up on pizza!

Capelin, another of the main fisheries, also ends up as fertilizer. They feed on zooplankton. Alaska pollock is another major fishery that contributes little to the fish markets of North America and Europe. It is most familiar as the source of fake crab on salad bars. And it is shamelessly passed off as crab in the crab cakes of some restaurants, which points to how scarce real crab has become. Pollock also feed on zooplankton, small invertebrates near the bottom, and on some capelin.

Pollock was the fishery in which Wayne Perryman had extensive experience. One cloudless morning, glorious for being in the shelter of an island off Baja California—we had suffered several days of 30-knot winds and 4-m seas—I asked him about his experiences.

"The fish deteriorate so quickly that they become mush. So, the main market is Japan where it is converted to the fish cake that accompanies noodle dishes. It also goes to Korea and Russia.

"The fish occur in dense aggregations and are caught in midwater trawls. A large net is pulled through the water, with floats to keep the top end up and weights to keep the bottom down. A typical set could pull up 10 tons of fish. The fish were caught by small U.S. ships, but then would be taken directly to large processing ships from other countries.

"The fishing fleet was large and had a high capacity to catch the fish. The boats had once harvested valuable king crab. With its collapse, they were ready to move to pollock and harvest it hard."

I was surprised when he told me that the fishery was year round; the Bering Sea and the waters between Kodiak and the Alaskan mainland, I knew from experience, could have bad weather even in midsummer. "Storms blew in unexpectedly, reached 65 knots (about 100 km/h), and boats would ice up and topple over. Even during the best weather a lot of vessels in the Shelikof Strait jostled for position, trying to get to the processing ship to unload their catches. It was hard to maneuver safely. During the long nights the water looked like a city, with all the lights sparkling on the boats."

Wayne's duties had also taken him to the open waters of the tropical Pacific to observe the tuna fishery there. Most yellowfin tuna were caught in the eastern tropical Pacific, and indeed that is why we were there; it was a far cry from the pollock fishery of the Alaskan shelf waters.

Tropical tuna are warm-blooded, superbly adapted to blue waters. (Bluefin tuna keep their body temperatures between 28°C and 33°C.) They swim long distances and eat voraciously when they finally find food. Tuna use their speed, up to 90 km/h, to overtake their fast-swimming fish prey.[6] Once they have left the food patch where they were hatched, their lives are high-speed, frantic journeys from one food patch to the next. If they fail to reach a new patch of food in time, they starve.

Tuna are predators' predators. They feed on fish, some of which feed on zooplankton, some of which feed on other fish that feed on zooplankton. Other large predatory fish include sailfish that hunt in packs like wolves. Sailfish herd their prey into tight schools as they flash their sails to scare the smaller fish. Then the predators charge through the densely packed school snatching food as they go.

The eastern tropical Pacific is one of the most productive tropical oceans. Nonetheless, the area is huge, the predators' prey is patchy and ephemeral, and thus finding tuna and other predators is no small task. However, the fishing technology was up to it.

"I had been seconded to the *Gina Anne* to observe their operations," Wayne told Lisa Ballance and me as we stood on the flying bridge drinking coffee one morning. "She was a beautiful boat, 60 m long. We left San Diego and landed in Acapulco 60 days later." On deck, observers with high-powered binoculars would search for flocks of seabirds—boobies and shearwaters—and schools of dolphins. Both were associated with tuna. Seabirds would also be detected by the ship's radar. "Each boat carried a Bell 47 helicopter whose job was to search even larger areas for dolphin schools."

The key to the success of the yellowfin tuna fishery in the eastern tropical Pacific is their association with dolphins. Although the two are found worldwide, nowhere else do they associate in surface schools the way they do here. "No one knows why this association is so prevalent here. There are a lot of ideas but no real answers." Lisa explained.

"One possibility is that the thermocline is shallow in the eastern tropical Pacific." The *thermocline* is where the temperature of water changes suddenly from warm surface waters to cooler deeper waters. "Nutrients in the tropics are generally found below the thermocline. If the thermocline is below the zone where it gets too dark, phytoplankton cannot use them. When the thermocline is shallow, the phytoplankton have access to light and nutrients, and hence their production is high." "Yet another complication in the story of how much plant production is in the oceans," I thought.

Lisa continued: "Anyway, there is some thought that tuna will forage down to the thermocline, but not below it because it gets too cold for them. Elsewhere, tuna are found too deep for them to associate with dolphins. The tuna that do associate with dolphins are large, as big as the dolphins themselves. This makes the fishery even more successful because of the large size of each fish.

"We know why the birds associate with these aggregations. The predatory fish chase their prey to the surface. Seabirds take advantage of the fact that the predators prevent prey from escaping to depth and they flock over the predators."

The *David Starr Jordan* carried much the same equipment and had much the same purpose: to find seabirds and schools of spinner and spotted dolphins. Wayne, on retiring from the NOAA Corps, now worked as a fishery biologist, photographing the dolphin schools from the ship's helicopter. From the photographs he could count the number of dolphins and measure their size, including the number of young that were with them. The preoccupation with dolphins was that tuna boats caught their fish underneath dolphins—"on porpoise" went the bad pun.

Wayne continued with his story. "When a school was sighted, the *Gina Anne* would come within 3 km and then launch its speedboats." Like sheepdogs, the boats would circle the dolphins into a tight group, rounding up strays and preventing animals from escaping. Once the roundup was accomplished, the ship would launch the seine skiff, which held the far end of the net. The *Gina Anne* would then encircle the dolphins, playing out the net until ship and skiff would join. At this point, the ship would pull in the purse string at the bottom of the net, preventing tuna and dolphin from escaping." Lisa added, "Although no one knows why, the tuna stick with the dolphins through this chase period. This bond is what makes the fishery successful."

Now the top string of the net would be pulled in. The problem was to get rid of the dolphins. This is a practical matter and has nothing to do with the ethics of this type of fishing. Handling large numbers of 100-kg, unwanted, dead dolphins on deck was a huge inconvenience.

Wayne was now drawing pictures for me. "The ship would draw the circular net into a long tube. At the end farthest from the ship was a special panel of fine netting, too fine to entangle the dolphins' beaks. The dolphins were frantically swimming from one end of the trap to another. But when they got to the far end, the ship would back up and the net would sink below the water and dolphins, but not below the tuna. The net would always remain deeper than the dolphins, which could then swim free. That process might be repeated several times before all the dolphins were released."

Once that was done, the now small net, full of thrashing tuna, was brought close to the ship. A dip net would hoist the fish from the ocean onto a brailler where unwanted fish such as sharks would be discarded. The tuna—mostly yellowfin, but also some big eyes—would be dumped into cold wells of salt and seawater below decks. The temperature was not enough to freeze the fish rock hard, but it made them hard enough that they were not easily bruised as they were jostled in the hold.

"All that was the ideal. One set could land 5 to 10 tons of fish, and the ship had a capacity for about 1200 tons. Two sets per day were possible, three on a good one."

"So what happened when it wasn't ideal?" I asked.

Wayne thought purse seining to be "an amazing and technically skilled operation, difficult and tricky." So from time to time there were "disaster sets." Sometimes a canopy formed in the closing net and dolphins would get stuck under it and drown. The fishers would jump into the water and pull the animals from underneath the canopy to free them. "A very scary job." Wayne continued, "In the dark it was easy to become disoriented, and there could be large sharks in the net, too."

Not all tuna were caught this way. One of the sights that had surprised me on the trip was the fish that would accompany even small bits of flotsam. An empty plastic bottle might have 100 small fish in attendance, easily visible in the clear waters. Larger items, such as logs, drew schools of large tuna, mainly skipjack and young yellowfin. On Wayne's trip the crew had discovered a tree trunk and set the net around it. They had netted 100 tons the first try, 80 and 60 on the next two, then passed it on to another ship that caught a steady 40 tons a day for 5 days.

Particularly offshore, where there was little debris, fish aggregation devices (as they were called in the trade) could exert a powerful attraction for the smaller tuna. Boats would drop a sheet of plywood overboard with a radio transmitter on it, let it sit overnight, then locate it the next morning and fish it.

The sheer ingenuity of catching fast-moving predators in tight groups scattered across such huge distances is surely the most impressive way we obtain any of our foods. However, it certainly isn't the only way of catching such fish.

Rather more mundane are long-liners, boats that put tens of kilometers of line over the side with thousands of baited hooks. We had seen a long-liner's line and floats even though the ship itself was well beyond the horizon. It was pursuing the largest tuna, bluefins, that could weigh three-quarters of a ton and fetch $80,000 (for one fish!) at market and over twice that in restaurants. What long-liners actually caught was a different matter: sharks, sailfish, marlin and other bill fish, seabirds, whatever would be attracted to the bait.

Not all fishing is in the water column; trawlers also go after flatfish and other bottom-living fish. The most widely used bottom fishing gear is the otter trawl. Its forward motion spreads a pair of otter boards (which can weigh several tons) that hold the trawl mouth open. The bottom of the trawl mouth is a rope that holds heavy steel weights to keep the trawl on the bottom.

This gear crushes, buries, or exposes marine animals while greatly changing the bottom of the ocean floor. Photographs of the bottom after trawling show that otter trawls act like giant ploughs whose furrows are clearly visible. A wide range of invertebrates share this environment, including corals, mollusks, sponges, sea urchins, and worms that live on rocks or pebbles on the seabed. Other animals, such as polycheate worms, burrow into the seabed. This environment is also home to many commercially important, but still immature fish.

The extent of this damage is extraordinary since it creates a disturbance much larger than the continuing massive clearing of the world's tropical forests. It is not well known because it remains out of sight beneath the oceans' surface. Les Watling of the University of Maine and Elliott Norse of the Marine Conservation Biology Institute estimated that bottom trawling ploughs 15 million km^2 of the world's sea floor each year.[7] That is an area directly comparable to the world's croplands, an area concentrated in the 30 million km^2 of nutrient-rich continental shelf. On average, the ocean bottom of these productive waters is trawled roughly every 2 years. That average hides the fact that a few areas escape trawling altogether, while others may be trawled 5 or even 50 times a year.

Almost certainly, the majority of the world's fishers are not engaged in large, commercialized fisheries. Across much of the developing world, fishers set out in small, leaky boats with ratty nets or improvised lines. The fish they catch are small—sometimes only a few centimeters long—and are whatever gets caught. Eaten fresh or dried whole and sold by the kilo in the local markets, they are an important source of protein in poor countries.[8]

From these few examples an unmistakable picture of human impact begins to emerge. To land-based visitors, the seas appear to be a vast, untouched wilderness. That is a mistake. Fishers go everywhere, from the pack ice of the Arctic to the pack ice of Antarctica, in winter and in summer. The conditions are often appalling and dangerous. They fish abundant, coastal fisheries, often with simple techniques, and they fish in open blue oceans with sophisticated technology to find their widely scattered prey. They also waste large amounts of fish and other animals, and they physically damage the ocean floor and coral reefs.

Visits to the local fish market took on a new significance following my return from the cruise. I had a greater curiosity to know where the fish had come from and a deeper appreciation of the difficulties in catching them. Above all, the disappearance of fish from the markets' displays tells a powerful story about how humanity has exploited and then depleted the resources of even the most distant and dangerous seas.

The Fishmonger's Slab

As for the land, my story finds a common currency in how much biomass we extract from the oceans. It is an incomplete story because it overlooks physical damage, among other things. Nor does it account for how we diminish the variety of life in the oceans. But it is a start.

Together, the top 15 fisheries account for about 40 percent of the world's total catch of about 95 million tons, or 23 million tons dry weight. The world's oceans produce over 100 billion tons of biomass dry weight per year, so the total catch is a tiny fraction, 1/20th of 1 percent, of what we might harvest.

Here is the complication. On land, most of our food comes from plants, not animals. From the oceans, plants contribute little to the total harvest, less than 2 million tons. That includes the seaweed wrapped around sushi in Japanese restaurants. Plants are likely to remain a small harvest: Filtering and eating phytoplankton is a prospect that is neither appetizing nor feasible.

How different are the natural histories of the animals we eat from the sea! Anchovies are the oceanic equivalent of cows and sheep, mostly filtering the phytoplankton from the water. Pollock and capelin are planktivo-rous; that is, they feed mostly on zooplankton. Pollock also eat capelin. Tuna are piscivorous, feeding on fish, some of which feed on zooplankton, while others are themselves piscivorous.

Each of these stages—phytoplankton, zooplankton, fish that eat the zooplankton, fish that eat zooplankton-eating fish, and so on is called a *trophic level.* It is the range in trophic levels from plants to sharks and tuna that causes the problem. How much of the oceans' productivity we con-sume cannot be represented simply by the catch itself. It depends on its trophic level.

Remember Jonathan Swift's famous doggerel about fleas?

So, naturalists observe, a flea
Has smaller fleas that on him prey;
And these have smaller still to bite 'em;
And so proceed ad infinitum.

This, too, is about trophic levels, though Swift was using it to criticize how lesser poets borrowed from greater ones, in much the way that movie sequels are always less engaging than the originals. Higher trophic levels have less biomass than lower ones, too. There are fewer calories to be had at progressively higher trophic levels.

The reason is that no conversion can be 100 percent efficient. A million calories of phytoplankton production eaten by zooplankton do not produce anything like a million calories of baby zooplankton. Some will be lost as waste in feces, and some will be lost in the energy burned by these small ani-mals as they go about their daily lives. The same principles apply to the fish that eat zooplankton, to the fish that eat those fish, and to us. By far the greatest fraction of the 2500 calories we consume each day goes to keeping us warm and providing the energy for our activities, and not (thank good-ness!) to growing our waistlines.

The fraction of the calories in the consumed phytoplankton that gener-ates zooplankton growth is called the *efficiency.* Compared to birds and mammals, which burn lots of calories to maintain a near-constant body tem-perature, the daily needs of zooplankton and fish are much smaller. Nonetheless, even for them only about 10 percent of the energy they con-sume goes to new growth.

This is the idea behind Pauly and Christensen's calculations. "Ten per-cent" means that a ton of phytoplankton production will yield only 100 kg of zooplankton production. The planktivorous fish that consume this zooplankton production in turn will produce only 10 kg of fish, and the piscivorous fish that eat them will produce only 1 kg of fish.

Now reverse the argument. Next time you eat a delicious swordfish steak, say a quarter of a kilogram dripping in butter, give a thought to the quarter of a ton of phytoplankton whose deaths by zooplankton made it possible. How quickly our human activities change nature! When I first wrote this paragraph, swordfish was on restaurant menus. It is now so badly overharvested that responsible chefs refuse to serve it (and responsible diners refuse to order it).[9] And a currently popular movie, *The Perfect Storm*, shows both the long-line methods to catch these fish and the fatal consequences to the fishers as they deplete their numbers.

This reverse logic would allow us to take the entire world catch of swordfish and calculate the total production of phytoplankton needed to support it. Now, take the comparable calculations for all the world's fisheries—at least 1000 fish stocks—and add up the numbers. (A fish *stock* is a kind of fish in a given place. For instance, the bluefin tuna in the Atlantic may have little contact with the bluefin tuna in the Pacific. Their fates may be independent of each other, and so fisheries biologists treat them separately.)

Adding the numbers over all the stocks tells us how much of the ocean's phytoplankton production we need to support our fisheries. This is the recipe that Pauly and Christensen employed, but there are some necessary complications.

The simple calculation for swordfish would work provided, first, that the efficiency is really 10 percent and, second, that a simple chain of eating phytoplankton to zooplankton to small fish to swordfish is a good description of what happens in nature. If the efficiency figure were really 5 percent, the same quarter of a ton of phytoplankton production would yield only 12.5 kg of zooplankton production. The planktivorous fish that consume those zooplankton would produce only 0.625 kg (625 g) of fish, and the swordfish that eat them would produce only 31 g (just over 1 oz) of swordfish. (This sounds like one of those nouvelle cuisine restaurants owned by a celebrity chef, with whom one negotiates reservations 2 months in advance and leaves the restaurant hours later, much poorer and still hungry.) Consequently, Pauly and Christensen had to justify why 10 percent is the right value for the efficiency of energy conversion.

Second, fish are like human beings: They do not get all their food from one trophic level. Vegetarians excepted, we eat fish with our chips and meat with our potatoes. There are lots of fish examples: Pollock eat both capelin and zooplankton.

Pauly and Christensen's solution to both problems was to work out the diagram of who eats whom—the food web—and by how much. For the small stuff, this is easy. Water samples contain zooplankton and phytoplankton. Put them in a shallow dish under a microscope and observe. Fish are too large for this approach. Fortunately, fish biologists are inordinately fascinated by fish stomachs. For commercial fish we have an endless supply. On oceanographic vessels, collecting tuna stomachs is an immensely popular task. Our cook on the *David Starr Jordan* was always on hand to turn the rest of the carcass into sashimi. Inside the fish stomach is a detailed and quantifiable record of what the fish has eaten.

Nonetheless, you have to want to know every last fact about fish to have all the necessary information. Daniel Pauly is an extraordinary compiler of fish facts. His knowledge is as broadly based as his life. A Frenchman educated in Germany and employed in the Philippines for decades, he is now at the Fisheries Centre at the University of British Columbia, Canada. He collects data sets passionately by the sheer force of his charm and enthusiasm and in all the languages that his travels have led him to acquire. One cannot imagine him slowing down until every last fact about every last kind of fish lies at his, and everyone else's, fingertips. It is precisely that passion that made the analyses of human impacts on the oceans possible. Only he could assemble the myriad pieces of such a puzzle. The collection is called Fish-Base, and it contains information on 80 percent of the more than 20,000 kinds of fish.[10]

To get the numbers of who eats whom, by how much, and with what efficiency, Pauly and Christensen examined 48 detailed case studies and obtained 140 estimates of efficiencies. Their average was close to 10 percent, and nearly two-thirds of the values were between 6 and 14 percent.

The second part of their work was to calculate an average trophic position for each type of fish. Pauly and Christensen estimated that tuna and some similar fish get 80 percent of their food from the third trophic level, which is fish that eat zooplankton that eat phytoplankton. The remaining 20 percent comes from the fourth trophic level: the fish that eat fish that eat zooplankton that eat phytoplankton.

The final stages of the process were already done by the FAO. Their fishery statistics group all the 1000 or so kinds of fish that we catch into 37 major categories on the basis of their natural histories. One of these categories includes bonito, wahoo, mackerel, tuna, sailfish, marlin, and swordfish. Some of the catches are large (skipjack tuna and yellowfin tuna, for example, account for about 3 million tons of catch per year). Others in this grouping have landings only a thousandth of this size.

The fishery statistics also group these catches into geographic regions, for example, the eastern tropical Pacific, the western tropical Pacific, and so on. The FAO's statistics are more detailed than the six major regions described in Chapter 8, but each geographic region can be placed in the right pigeonhole. The catches of yellowfin tuna, for instance, come almost entirely from the eastern and western tropical Pacific and the tropical Indian Ocean. These are all tropical open ocean systems.

We can now insert the numbers into the equation I described earlier. Some 0.73 million metric tons (dry weight) of tuna and similar fish (including bonito, wahoo, mackerel, sailfish, marlin, and swordfish) are caught in open ocean systems. Another 0.30 million tons dry weight come from tropical shelves.

Suppose that all these fish fed at exactly trophic level four:

0.73 million tons of bonito, wahoo, mackerel, tuna, sailfish, marlin, and swordfish would eat:

7.3 million tons of smaller fish, which eat:

73 million tons of zooplankton, which eat:

730 million, or 0.73 billion, tons of phytoplankton from the open ocean.

Pauly and Christensen's databases showed that this is too simple a formula, because these fish derive 20 percent of their food from the next higher level. So, the previous numbers are 80 percent correct (which comes to 0.58 billion tons of phytoplankton). To that we must add 20 percent of 7.3 billion tons of phytoplankton. If these fish got all their food at this higher level, then 7.3 billion tons of phytoplankton production would be needed to support the 0.73 billion tons of zooplankton production, the 73 million tons of smaller fish, the 7.3 million tons of predatory fish, and the 0.73 million tons of tuna. This means that it takes close to 2 billion tons of phytoplankton production to support our consumption of swordfish steaks, tuna sashimi, and tuna salads, and all those other great predatory fish that taste so wonderful.

As it happens, these fish are the only ones we harvest from the open ocean. Compare these 2 billion tons of phytoplankton needed to support these fisheries with the 89 billion tons of phytoplankton that the oceans produce annually and the number is quite small.

One more example: the anchoveta, this one from an upwelling. The total catch of this fish alone is three times that of the tuna and similar fish combined, and more in good years. Yet the anchoveta feed on phytoplankton and zooplankton, a much lower trophic level than the tuna. As a consequence, the anchoveta require only one-tenth the amount of phytoplankton production to support it, even though the fishery is much larger. But upwellings occupy an area of much less than 1 percent of the open ocean, and hence the catch is more concentrated.

What Does Not Get to the Fishmonger's Slab

One final element is missing from these calculations: the *bycatch*. These are unwanted fish, but also birds, mammals, turtles, and invertebrates that are caught incidentally. Examples include dolphins that drown during purse seining operations and unwanted fish caught on long lines. (Moviegoers may recall the large shark hauled on board in *The Perfect Storm*.)

How large is the bycatch? Pauly and Christensen estimate 27 million tons, worldwide, annually. This is about a third the size of the commercial catch. Possibly it is much too low a figure, especially when one considers the detailed estimates from wasteful fisheries like shrimping. Some global totals are as high as 40 million tons a year. The figures are not included in the fishery statistics.

The shrimp boats that trawl offshore have a high fraction of bycatch.[11] Sponges, jellyfish, soft corals, crabs, squids, sand dollars, and starfish are swept into the funnel-shaped nets. Our impact on them is surely large but no one measures mere invertebrates. They are not the only victims. Before the introduction of turtle excluder devices (or TEDs), between 5000 and 50,000 sea turtles were killed each year off the coasts of the United States alone.

Fish are killed too, and the shrimp boats off the coast of Texas and elsewhere in the Gulf of Mexico have severely reduced the populations of commercially valuable red snapper. Several studies suggest that the bycatch of fish from the world's total catch of shrimp of roughly 2.5 million tons is between 5 and 19 million tons.[12] These numbers are from three to seven times the size of the catch, not the one-third that Pauly and Christensen estimate for the global average.

Trawling for shrimp and fish is not the only method that destroys large numbers of unwanted animals. Gill netting is the other common method of fishing. Hundreds of kinds of nontarget animals are caught, including seabirds, turtles, and mammals. The Japanese squid fishery is "credited" with reducing the populations of the northern right whale dolphin, the Pacific white-sided dolphin, and the commercially important albacore tuna. The mortality among marine mammals is substantial. The few available data suggest kills of marine mammals on the order of a few for every time the nets are set. The Japanese squid fishery sets gill nets about 50,000 times per year. Kills of dolphins and porpoises probably number in the hundreds of thousands.[13]

As a fraction of the total harvest of biomass from the oceans, these marine mammal kills seem insignificant. This view assumes that all that matters is biomass, that we count a kilogram of an endangered whale the same as a kilogram of anchoveta. For the purposes of this chapter we do. In later chapters I take a different view.

Wayne Perryman had first-hand experience of bycatch from the purse seines. "As the tuna were dropped on deck, it was a frantic scramble to get the tuna into the hold and separate out what wasn't wanted. It had to be done quickly or else the tuna would die in the net, sink to its bottom, and their weight would destroy it. Black marlin, other billfish, and white-tip sharks were sorted out and thrown back on deck. By the time all the tuna were in the hold, there would be a wall of dead and dying large fish to be tossed back into the water. There were countless numbers of smaller fish, including tuna too small to go to market."

His estimate of about a third matches the world average. Nonetheless, bycatch estimates are difficult to make, and they are extremely controversial. Kills of marine birds, and especially mammals and turtles, are highly emotional issues. There is no incentive for fishers to report such kills and every incentive to keep quiet about them. A few boats carry observers to record the numbers. One of my shipmates on the *David Starr Jordan* had done this. She told me that observers on long-liners would report many dead billfish and sharks per set, while even the cooperative captains willing to record bycatch would report just one. Pitman had done some scientific long-lining to check on the accuracy of bycatch estimates. The crew had caught a few sharks, and every single one had a hook in its mouth; they had all been caught previously and were lucky to have survived.

The Oceans' Spreadsheet

Pauly and Christensen's work estimated the phytoplankton production that is needed to support the bycatch for the fisheries in different regions in exactly the same way as for the catch itself. For open oceans, for instance, that came to about 10 percent of the catch.

In summary, these were their results.

- Fisheries of the open ocean need about 2 billion tons, and their bycatch 0.23 billion tons of phytoplankton production to support them; combined, this comes to about 2 percent of the 89 billion tons of annual phytoplankton production of the open oceans.
- Fisheries of upwellings need 0.72 billion tons, and their bycatch 0.11 tons; combined, this is 28 percent of the 2.9 billion tons of annual production of upwellings.
- Fisheries of tropical shelves need 1.08 billion tons, and their bycatch 0.1 tons; combined, this is 24 percent of the 5.9 billion tons of annual production.
- Fisheries of nontropical shelves need 3.08 billion tons, and their bycatch 1.35 tons; combined, this is 35 percent of the 12.5 billion tons of annual production.
- Fisheries of coastal waters and reefs need 0.17 billion tons, and their bycatch 0.05 tons; combined, this is 6 percent of the 4 billion tons of annual production.
- Fisheries of rivers and lakes need 0.3 billion tons; no estimate of bycatch was available; the catch alone is 23 percent of the 1.3 billion tons of annual production.

Averaged across all the oceans, only about 8 percent of the total phytoplankton production goes to support the fisheries we consume. This average is quite misleading, however. Across the open oceans only 2 percent of the productivity supports our fisheries. Yet across the geographically restricted, highly productive parts of the oceans and freshwater lakes, between a quarter and a third of all plant production supports our fisheries.

Green Slime Sandwiches, Anyone?

Can't we simply feed at a lower trophic level? For example, cod feed almost as high on the food chain as do tuna. By feeding so high on the cod (sorry!), we are energetically profligate. Wouldn't we do better to feed on what the cod eat?

In a paper published by *Science* in 1998, Pauly and Christensen explained how fisheries have changed over the last few decades.[14] They found that although the total fish catch has remained rather constant over this time, we are now feeding on lower trophic levels than in the past. That is because we have driven the stocks of many of the valuable, top-predator fish (such as cod, tuna, and swordfish) to low levels. To replace these fisheries, we are moving to lower trophic levels where the fish, though more abundant, are less appetizing.

There are limits. If we harvested phytoplankton and not fish, the biomass we extract from the oceans would be a negligible one-twentieth of 1 percent. Biomass is not everything. Green slime sandwiches are not likely to catch on. It's not just a matter of some clever advertising campaign that would encourage us to use phytoplankton spread on toast instead of Vegemite. Phytoplankton may not be edible, just like the majority of stems and leaves of terrestrial plants. Plants on land and in the oceans have few ways of avoiding being eaten other than making themselves tough, prickly, bad-tasting, and low in the nutrition that other living things need. Even if they were edible, much effort and energy are required to filter the seawater to get even large phytoplankton.

We are no more likely to eat zooplankton than we are to eat phytoplankton. Apart from small harvests of seaweed, and near-shore harvests of mollusks, the lowest trophic level we consume are fish that feed on at least some zooplankton. We feed almost two trophic levels up the food chain.

Even at this trophic level most of what we catch goes to fishmeal. My preference for anchovies on pizza is not widely shared. (In 1996 Domino's Pizza reported that squid is their most popular topping in Japan, tuna in England, and mussels and clams in Chile.[15]) At low trophic levels, the market price—for most fish, less than $100 per metric ton—restricts the harvest to the abundant, concentrated fishes that can be harvested efficiently.

Just what is abundant in the world's oceans? Elliott Norse in his important and highly readable guide to the oceans, *Global Marine Biological Diversity*, introduces the genus *Cyclothone*.[16] These fish are perhaps the world's most numerous and widely distributed. Yet not one kind appears in the FAO yearbook of fishery statistics in its 1000-entry list.[17] It is not just their ugly name. I cannot imagine a waiter asking, "Does Sir want his bristlemouth almondine or à la meunière?" Bristlemouths are small, made up mostly of water, and do not occur in the schools that would make their harvest practical.

Not that some nations don't try. Consider the abundant shrimplike (and largely herbivorous) Antarctic krill.[18] They presented a tempting target to the Soviets. The mid-1980s harvest was over 400,000 tons. The Soviets developed advanced technology to harvest krill, but the costs of the final product were high. And the product? When was the last time you ate krill? In the high latitudes, the fishing season is a short 150 days. When krill are feeding they have an unpleasant "grassy" taste and spoil quickly. After they are caught they must be peeled quickly or unacceptably high levels of fluoride leach into the meat from the shells. The first commercial product, "Okean," was made from coagulated protein, but the most valuable products are the peeled tails, especially of the ripe females. These have the highest fat content and the best flavor. However, the machines that peel them damage and waste over three-quarters of them. The harvest dropped dramatically with the breakup of its principal exploiter, the Soviet fishing fleet. Post-Soviet Russians wouldn't eat krill either.

Can't we take more from the open oceans? Only about 1 percent of our total global catch comes from the open oceans. And only 2 percent of the phytoplankton production goes to support the catch and bycatch.

In these low productive ecosystems we feed on tuna and similar fish that are four trophic levels up the food chain. In the oceans, we feed at paradoxically higher trophic levels in the least productive open oceans. Why? Again, there is a parallel to the land. In the terrestrial least productive ecosystems, deserts and tundras, we use "specialized collection systems" (goats, camels, reindeer) to harvest the primary production for ourselves rather than exploit plant life directly. We use tuna (and goats, camels, and reindeer) to concentrate food because we cannot efficiently do it ourselves.

In open oceans, as in deserts, resources are widely scattered and unpredictable. The goats mentioned in Chapter 5 were able to eat only a small fraction of Afghanistan's plant productivity. Most of that production comes in a short burst after the winter rains. The rains fall sporadically and locally and vary greatly from year to year, while the goats must eat year-round and year after year.

The plant production of the open oceans is similar. A localized phytoplankton bloom may be quickly devoured by the zooplankton. Will the fish that feed on them find this patch? Often they will not. Pauly and Christensen suggest that zooplankton-eating fish may consume only 20 to 25 percent of the zooplankton. In turn, will the tuna roaming the desertlike expanses of open oceans find these fish? Patchy resources make it impossible for tuna to take more than a small percentage of what is available to them as they rush from one food patch to the next.

As Pauly and Christensen put it, "The prospects for increases (in catches) are dim." The only consistently increasing fisheries are of carp, tilapia, and farmed salmon. Their catches more than doubled in the decade from 1984 to 1993. They now represent nearly a fifth of the world's fish consumption.[19]

Freedom of the Seas

Huxley concluded his inaugural address to the Fisheries Exhibition with a plea to oppose the fishery legislation that opinion demanded. It was a punishment "on the simple man of the people, earning a scanty livelihood by hard toil." It was the man who made such unnecessary laws who deserved the heavier punishment, he urged. The fisherman, like his fathers before him, should have unfettered access to the oceans. He was asking for freedom of the seas.

This ought to have been the right answer. Two-thirds of the planet's surface is ocean. The oceans have huge fish stocks that are hard to extract. Yet Huxley was spectacularly wrong. The blue open oceans constitute 90 percent of the planet's saltwater, but only 1 percent of the global fish catch comes from them. From the remaining 10 percent—the productive green seas—we already require between a quarter and a third of the plant production to support almost the entirety of our wild-caught fish. Those fractions are broadly similar to humanity's take from the land.

The world's fisheries cover all the seas, coastal to midocean and from sea ice to tropics, regardless of danger and difficulty. No large, untapped fisheries have been overlooked. Humanity's total impact—our numbers, our technologies, and our burgeoning affluence—has taken us to remote and inaccessible parts of our planet.

Taking a quarter to a third of the oceans' production may not seem like much, but it may already be too much. Even at current levels of exploitation, let alone levels that will accompany a doubling of the human population, we are already harming our natural resources. The evidence shows that we are already fishing lower trophic levels, even though those fish are less valuable. This fact alone suggests that all is not well.

Huxley got his wish; many fisheries were only loosely regulated. One consequence was 400 angry fishers ramming and kicking in the doors to a conference room in Newfoundland, Canada. In one of Huxley's inexhaustible fisheries, so few fish remained that it was to be closed down.

10

The Wisdom to Use
Nature's Resources

St. John's, Newfoundland, Canada, 2 July, 1992: John Crosbie, Canada's
Minister of Fisheries and Oceans, announces at a news conference that
the great cod fishery of the Grand Banks and the Labrador coast will close
for 2 years. Nineteen thousand fishers and fish plant workers will be out of
work in a province where 22 percent are unemployed already. The 400 fishers who have come to hear Crosbie in person are directed to a ballroom to
watch a prerecorded video. As *Maclean's* magazine recorded, they quickly
decided differently.[1] Crosbie's aides try to keep them at bay, while frantically calling for police help. A police escort has to take him past the angry
fishers to his car. "There's no home for you in Newfoundland," someone in
the crowd shouted at him. Crosbie is a Newfoundlander.

Cod was the economic basis for the settlement here in the fifteenth
century.[2] Fishing cod gave modest prosperity to hundreds of small, isolated
coastal communities. The houses, I thought, looked neat and comfortable,
their colors brightening the often foggy days. For most of the twentieth century, the total cod catch had been about 200,000 tons. Some 800,000 tons
of cod were caught in 1968. Even the 1980s were good years. As late as
1988 the quota—the limit on how much could be caught—was set well
above 300,000 tons. But by 1991 only 70 percent of the 185,000-ton quota
was realized, and future predictions looked worse. Crosbie reduced the
quota, then closed the fishery altogether.

Young people in the villages were already leaving for better futures elsewhere and wondering whether they, and the cod, would ever return. The
older generations were asking what else they could do but collect government handouts? Over a 2-year period those benefits would cost half a billion dollars. Suggesting that he might resign if his cabinet colleagues did not

improve on that offer, Crosbie eventually proffered benefits of nearly twice that amount.

The news continued to worsen. By 1995, some 35,000 people were directly affected by the closure, and many more were affected indirectly. Cod numbers declined further in 1994, leading to discussions of the fishery's being closed for a decade or more. The costs to the government rose again.

How could this have happened? The countries that exploit fisheries in northern latitudes are rich, industrialized nations, with centuries of fishing experience and with excellent scientists devoted to studying the fisheries. Moreover, the loss of fisheries to the economies around the North Atlantic and North Pacific are not merely recent events. We have 200 years of experience of pushing fish stocks to (or over) the brink of extinction.

This chapter asks a simple, if alarming, question: Why do we destroy what we should keep? Although fishery examples predominate in this chapter, much more is at stake. Something is terribly amiss when we destroy resources on which we depend. If we cannot manage fisheries properly, what hope do we have for managing all our other natural resources, including our rapidly disappearing tropical forests, and our degrading drylands?

How is it possible to destroy resources in the ocean or on land when we take such small fractions of their annual production? How is it possible to destroy ocean resources when the oceans are so vast? On average, only 35 percent of the plant production in seas like the Grand Banks supports the fish we eat. Could a percentage this low still be too high to be sustainable? If so, how close are the world's other fisheries to being exhausted? We should ask similar questions about vast tropical and boreal forests, and marginal drylands.

The answers to this chapter's simple question are complex, and we will draw connections to answers in earlier chapters about the land and freshwater. In providing these general answers, this chapter is a capstone for all preceding chapters. And it starts with a long history, one richly documented in fact and fiction: whaling.[3]

Right Whales, Wrong Whales, and Dead Whales

By 1800, whalers had nearly exterminated the northern Atlantic populations of right and bowhead whales. Right whales were "just right"; large and slow moving, they fed near the surface and often inshore, and they floated to the surface when harpooned. Whalers *did* exterminate the Atlantic population of the gray whale. Then they moved on to humpbacks and sperm whales. As these stocks declined, James Cook and others brought news of Botany Bay in Australia, of Hawai'i, and of Tahiti, then news of the Derwent River in Tasmania, where right whales gave birth in such large numbers that it was dangerous to cross the river in a small boat. From Nantucket, New Bedford, Mystic, Provincetown, Gloucester, and other villages along the New England coast, from Hull in England, and from Le Havre in France, whalers moved into the Pacific. By 1792, Captain Bligh on his second—and mutiny-free—visit to Tahiti was already complaining about them. English captains were chasing sperm whales off the coast of New South Wales after dropping off their passengers at the penal colony.

Whaling was intrinsically dangerous, but the wars between Britain and France had the additional risk of the loss of a ship or cargo after a long, successful voyage. The peace that followed Waterloo made the rewards greater. The first whalers arrived in Hawai'i in 1820; 100 were there in 1824. There were nearly 600 in Hawai'i in 1846, over 80 percent of the New England fleet. Each voyage averaged 100 whales per trip, and a voyage could last as long as 4 years.[4]

At the end of the 1800s the improved technology of steam engines over sail and of exploding harpoons over hand-thrown lances increased the harvest of faster-swimming whales—until then considered as the "wrong" whales. Whalers harvested nearly 30,000 blue whales in 1931. Thereafter the numbers declined year after year, with only the war years of the early 1940s to give them respite. By 1947, nearly 10,000 blue whales died; by 1963 the stocks were too low to support a catch. But no matter; whalers had begun to hunt the smaller, less valuable fin whales. In the 1930s, their catches were only a few thousand a year, but the numbers rose to over 25,000 in the early 1960s. The catch dropped precipitously in the mid-1960s, and to almost nothing by the mid-1970s. Then the whalers turned to the sei whale, which no one had bothered to hunt until the late 1950s. The sei whale catch peaked as that of the fin whales was crashing, only to crash itself a few years later. The harvest moved to the yet-smaller minke whales.[5]

There is only one way to interpret this story. Over two centuries, generations of whalers relentlessly exterminated first the easy and profitable whale stocks and then the more difficult to catch and less profitable ones. The consistent pattern had been the deliberate, rapid depletion of one whale species, then another, always to very low numbers and sometimes to extinction.

You may prefer the descriptions of whaling in Melville's *Moby Dick* or those in Patrick O'Brian's *The Far Side of the World*. Both impart a sense of something unworldly in the hunt of the planet's largest animals by men in small boats.[6] But the relentless destruction of one animal and then another is a very familiar story—just check your cookbooks.

Fish Recipes

I need only compare my two editions of *Joy of Cooking*.[7] The 1974 edition has 8 recipes for cod, 4 for haddock, 11 for herring, 5 for "roe," and only a few for other fish. By the 1997 edition, the recipes reflect the severe depletion of ranks of these once common fish. There is only one recipe for herring, for instance. A recipe has been added for orange roughy, unknown in 1974. In turn, Chilean sea bass has now replaced orange roughy (a long-lived, slow-growing fish of Pacific seamounts) as the delicate white fish of choice in restaurants. Cookbooks move through fish like whalers did through whales.

You insist on real numbers? FAO data show that after the large fisheries of upwellings (such as anchoveta), the next two largest groups are, first, the cods, hakes, plus haddocks and, second, the herrings plus sardines from the North Atlantic and North Pacific continental shelves. Together, these two groups once accounted for about 20 percent of the global catch. All are in decline. Like the Atlantic cod, Cape hakes (from the South Atlantic) are down by 60 percent from their peak catch and haddocks (from the North Atlantic) are down 40 percent from their peak.[8]

As an example of the sheer diversity of the overfishing problem, consider the United States. Congress requires the National Marine Fisheries Service (NMFS) to report annually on the status of all fisheries whose major stocks are within the country's economic exclusion zones. These fisheries come from the Atlantic, the Caribbean, the Gulf of Mexico, and the Pacific from off San Diego to the Bering Sea out to the west of the Hawaiian Island chain. Also included are fisheries of the western Pacific Trust Territories.

The 1999 report found that 98 fish stocks were overfished and 5 stocks were close to being so.[9] Another 127 stocks were not overfished. The conditions of 674 stocks were not known. Management efforts removed 10 stocks from the previous year's overfished list, though 18 more stocks were added. Of the 127 stocks, 47 were from the western Pacific and 30 from the North Pacific. For the major fisheries—ones in the Atlantic, the Gulf, and off the U.S. Pacific coastline—overfished stocks outnumber by 2 to 1 the stocks that are less depleted.

Lest you take solace in the condition of the nearly 700 stocks about which we don't know enough, most of them are poorly known because they are just not that important. One of them, however, is the barndoor skate (an appropriate name for this very large fish). It is caught as a by-catch of western North Atlantic fisheries. In a 1998 paper in *Science*, Canadian fisheries biologists Jill Casey and Ransom Myers make a compelling case that fishing is the cause of this fish's extinction from the region.[10] Too large to be overlooked, the numbers recorded fell progressively year after year until no one recorded them at all. Another is the black grouper, the grouper that fetches the highest price per kilogram at market. It is rare to get one in the Florida Keys these days. There the fishers are moving down the fish list, taking progressively less valuable groupers and snappers.

Why are these collapses so ubiquitous? Why can't we harvest a fishery—or a forest or grazing land—and sustainably reap the maximum economic reward decade after decade? There are several explanations.

The Tragedy of the Commons

In Huxley's day, the seas beyond a few kilometers offshore were international waters to which no nation claimed sovereignty. Even today, when countries have large economic exclusion zones, most of the world's oceans are outside them. We treat the oceans like a *common*, an area of land held jointly and shared alike by a community. It is land (or ocean) indivisible and unregulated, like many forests and grazing lands.

Garrett Hardin, in a now-classic paper in *Science* in 1968, explained the consequences in terms of grazing.[11] You choose to graze an extra cow on your village's common; you get the reward from the milk or meat. Damage to the common's pasture is shared by everyone. That is a deal you cannot refuse. But, then, neither can your neighbor to whom the same deal applies. All the incentives are to exploit the commons to the point of destruction— hence the tragedy. A commons system can work, but only if there are not enough cows to do any harm. The same principle works in the oceans. It can be a valuable commons only as long as there are few fishers.

The cod off Newfoundland were once fished by the coastal communities alone. In the 1960s, modernized boats allowed fishers from Spain, Portugal, and elsewhere to endure the ice and storms of Canadian winters. Foreign ships contributed to the peak harvest of 800,000 tons in 1968.[12]

Nations have long recognized the problem. Its resolution is politically popular and ecologically sensible. Extend the economic exclusion zone as far out to sea as your gunboats can effectively patrol. Take possession of the commons. Your fishers will vote for you, and the foreigners who can't vote (and who were stealing your fish) will be driven off. Your scientists can now manage the fishery (more votes, more applause). Environmentalists will support your statements about sustainable, nondestructive use of the environment. If the foreigners come back, the one-sided naval engagement will generate as much patriotic fervor as an international soccer match, but without the casualties.

Canada declared its 200-mile exclusion zone in 1977, but not all fish live within the protected zones. Cod also live outside the zone, where foreign trawlers fish. Canada's own massive offshore trawlers were simultaneously working the commons within that limit, while the inshore catch declined almost annually. Economic exclusion zones may turn an international commons into a national commons, but it is still a commons.

Open ocean fish such as bluefin tuna may live almost entirely outside such zones or move readily in and out.[13] Such fish are particularly vulnerable. For exactly the same reasons, so are many of the world's forests and grazing lands. Many nations do not treat the people of their rainforests as if they had ownership. The one who gets to the trees with a chainsaw first claims them.

Many people believe the selfish, free-for-all scramble that attends commons systems is the most important explanation for our inability to manage sustainably. Perhaps it is, but it is certainly not the only explanation.

The Freshman Phenomenon

It would seem reasonable that as a fish species becomes rarer, it becomes harder, and hence more expensive, to catch. It would also seem reasonable to then give depleted stocks a break and allow them to recover. There are two reasons why this doesn't work for many fisheries. First, many fish, bluefin tuna included, are an incidental catch. Long-liners would not purposely pursue the last of its kind with the passion of a Captain Ahab. Rather, they might catch the last one incidentally while going after smaller, less valuable tuna. This logic applies to large sharks caught the same way.

For the other explanation, you have to watch *The Freshman*. In this funny movie, a gullible first-year undergraduate is lured into the family of a Mafia don played by Marlon Brando. Combining the same mixture of overt kindness and veiled terror that characterized his role in *The Godfather*, Brando's character runs an international supper club. The entrée for each supper is the last one of its kind, so the admission is exclusive, and the price is beyond the wildest dreams of a five-star restaurant. When a valuable resource becomes rare, the price goes up, and then so do the incentives to acquire it. The resource becomes rarer and yet more valuable.

The don turns out to be a kind, sensitive guy after all. His celebrity chef does not slaughter the last one of its kind but substitutes smoked turkey. Hollywood always has happy endings. Nature does not.

In a paper in *Nature* published in 1994, Harvard ecologist Steve Palumbi and his colleagues investigate whale meat.[14] Raw and thinly sliced as sashimi, it is a prized delicacy in Japan. A few hundred minke whales are caught by Japanese whalers each year for "scientific purposes." (The science behind these activities eludes me.) Minke meat can be sold. Using DNA sequencing, Palumbi and colleagues found that not all whale meat for sale was minke. Some of it was from humpback whales, despite international agreements that nations not harvest them.

In a similar paper, a different group examined samples of caviar sold at a famous outlet in New York. The roe were not always from the fish indicated on the label. Sometimes they were from endangered Russian sturgeon.[15] The problem is a general one. Those who buy knickknacks made from elephant ivory and rhino horn are generally unwilling to accept plastic as a substitute. Patent medicines from tiger gallbladders and the like become more expensive as tigers decline, putting yet further pressure on the tigers.

The Discount-Rate Dilemma

Hollywood doesn't always supply the right characters, so I must now invent one. You own a fishery and you calculate exactly how much fish your fishers can catch per year, year after year, without exhausting your fishery. (You have graduated at the top of your class in fisheries ecology.) Suppose you own a million dollars worth of fish and can harvest 5 percent of them per year. This you proceed to do, living modestly, priding yourself on the fishers and their families whose livelihoods and communities you sustain, your commitment to sound environmental practices, and the rest of it. You are not this story's protagonist.

An observation nags you. There were only a few rows of fellow scientists at your university graduation, but half the auditorium was filled with graduates in business administration. One, our protagonist, also owns a fishery. You meet at an alumni dinner. Renewing your acquaintance, you ask:

"Now what is that fishery you are in—and how's business?"

When he replies, "Scarlet roughy," your heart sinks.

"Wasn't that the fishery that was driven to extinction only a few years ago?" you ask, hoping not to offend.

"Absolutely," he enthuses, "and now I am driving Namibian sea bass to extinction!"

His stories of his kids in exclusive private schools and his wife's difficulty in getting a reliable maid tell you that you've done something wrong. And indeed you have! Had you quickly harvested your million dollars worth of fish, driving them to extinction, you could have taken the money and invested it in almost any kind of bank account and gotten a better return on your money than the 5 percent your fishery returned. You could have spent part of this money by buying another fishery and driving it to extinction.

If you draw a moral from this story, would you draw a different one had I made you and your acquaintance a magnitude poorer? Now he barely supports his family and yours is always hungry. Many of the world's poor have to take short-term gains in order to live.

Canadian ecologist Colin Clark, from whose work I have extracted this story, makes the point with the requisite mathematics. Technically, we discount the value of the future. Any resource that does not generate a rate of return that exceeds what the money would yield if invested elsewhere doesn't have a future. You are better off liquidating it.[16]

Where should we look for such resources? The most likely candidates are the animals that grow the slowest, because they will return the lowest rate. Across the animal kingdom, a rough guide is that the larger the animal, the more slowly it will grow. The slowest-growing marine animals are the largest: the whales. Exactly as Colin Clark's economic arguments predicted, we harvested one whale stock after another to extinction, or close to it, and did so as quickly as possible.

The description works well for a whole slate of valuable fish that have appeared briefly in fish catch tables—and in cookbooks—and then disappeared. The different kinds of grouper have disappeared in a similarly predictable sequence from the Gulf of Mexico.

Among plants, the slowest growing are the large trees of old growth forests. Substitute old growth forests for whales in the previous few paragraphs and you have a partial explanation of the histories of forest clearings.

You protest that my chief character is fictitious. You're right. The actual protagonist, as Carl Safina describes in Song for the Blue Ocean, cut trees in the Pacific Northwest.[17] He bought a forestry operation that had sustained a community for decades, clear cut it, put the loggers out of work, pocketed a high profit, and returned to England. Safina does not say whether this individual, who was knighted, bought his knighthood with a significant payment to a political party, as was the rumor in Mrs. Thatcher's administration, but the story brings me to my next explanation.

Perverse Subsidies

Willie Sutton's famous retort as to why he robbed banks was "because that's where the money is." What a comedian! Banks aren't where the money is. After his last attempt at humor, he got 17 years. Our protagonist knows better.

"Of course, he wouldn't be where he is today were it not for the government's help!" you think to yourself about our leading character. He would not admit this. He is complaining about how the current government makes it hard for business people like him to make profits.

"All the regulations, all those greens with their concerns about the environment. And you wouldn't believe the taxes I have to pay!" He's been going on like this for almost an hour. "Why, when I had dinner last week with the *President** I gave him a piece of my mind." (*You may substitute whatever political office fits, since I mean this to be an equal opportunity criticism.)

You know that the dinner cost him a lot to attend and that this contribution was a small part of what he gave to the current administration's political party. But it was a smart business move. The government gave his company low interest loans to buy the boats, subsidized the fuel costs of running them, paid for the fisheries research, and provided other goodies. No wonder he has money and enough to spare to support politicians. The government is where the money is!

Since a politician is valuable only as long as he or she is in office, our protagonist always hedges his bets. His brother, an equal partner in the family company, gives generously to the other political party. Candidates of both parties need the money, because without it, they cannot afford to air TV commercials that are essential to their reelection.

You cannot have missed stories like these—they appear almost daily in election years—and you worry about their consequences, as indeed do responsible politicians themselves. In the context of using natural resources, however, you ask: "Shouldn't governments invest to further the public good?" Even those who think the answer should be "yes" generally expect the investment to turn a profit. Perverse subsidies are those that lose money.

Those losses are massive. Look at some global totals. FAO has produced a balance sheet for the world's fisheries.[18] In 1989 the total worth of the world catch was about $70 billion. Being a naïve fisheries student, you expect that the costs of this harvest will be less, so there will be a modest profit. Think again!

The annual cost of operating the global fishing fleet was about $92 billion in 1989. This sum consists of routine maintenance ($30 billion), fuel ($14 billion), insurance ($7 billion), supplies and gear ($18 billion), and labor costs ($23 billion). With these costs alone, the loss was $22 billion. The boats and equipment are worth about $320 billion, and a fraction of this—FAO suggests 10 percent or $32 billion—should be the annual cost of paying off the loans.

Worldwide, the fishing industry loses about $54 billion a year. No private enterprise could long withstand expenses that are nearly twice the receipts. Massive subsidies supported the fleet of the former Soviet Union. Its fisheries dropped dramatically following its breakup. Krill harvesting was one example.

Some specifics: The Japan Fisheries Association estimates the credit balance extended to its fisheries at $19 billion, noting that "the government financing system will assume their liabilities . . . [in the event of] financial difficulties." The European Union similarly supports the construction of new vessels and the modernization of old ones.

The supports come in a variety of direct and indirect ways. Nations may guarantee the price of the fish, subsidize the cost of the fuel, provide low-cost loans, or even give outright grants for new boats and equipment. Your college acquaintance may be using a boat that you, the taxpayer, helped to buy. Moreover, the government likely supports the ports where his boats land his fish.

Back to Newfoundland cod. The Canadian government has already paid billions of dollars in subsidies to sustain coastal communities that have few other opportunities for employment. The government ensures a supply of cheap protein. This keeps the voters happy because they do not realize that their taxes go to support it. Everyone assumes the supply will not run out—at least not while the current lawmakers are in power.

Agriculture receives similar subsidies but on a more massive scale. For instance, the U.S. government massively supports sugar prices rather than allowing free trade from other nations. Uncle Sam also pays for the construction of thousands of miles of dykes and levees that permit this thirsty crop to be irrigated. As a consequence, large parts of the Everglades ecosystem are damaged by runoff, which we, the (taxpaying) people, have to pay to clean up.

The U.S. Army Corps of Engineers and the Bureau of Reclamation spent tens of billions of dollars on over 1000 large dams in the western states. Sometimes, each project cost that much. As Marc Reisner explains in his compelling book, *Cadillac Desert*, many of these projects never had a prayer of supporting enough agriculture to pay for them. And they didn't![19]

Deforestation within the United States and throughout the world is supported in a similar way. In one of the most egregious examples, the Tongass National Forest of southeast Alaska gets only five cents back on every dollar it spends on cutting the trees and maintaining an extensive network of roads to make the cutting possible. Much deforestation in the Amazon was driven by tax incentives for those who clear the land of forest (and its indigenous peoples) and who put cattle in their place.[20]

Naturally, one's own subsidies are always essential; it would be unpatriotic to think otherwise. They are justified as a public good—we have to have fish, don't we? And fishers? And boat builders? And scientists to study the problems? And costly dams to provide cheap water that even at such low cost cannot be repaid by the agriculture it supports? And permits for grazing drylands that generate less rent than others would pay just to live there, so we can maintain a cowboy way of life? The enthusiasm politicians display for cutting costs always means someone else's subsidies, never those of their own constituents.

Norman Myers and Jennifer Kent estimate that subsidies total trillions of dollars worldwide per year.[21] The global sum of all economic activity is on the order of a few tens of trillions of dollars, by comparison.

How does all this relate to the impact of human beings on the planet? Ecologists have their own version of the economists' law of supply and demand. Fishers are predators, hunting wild prey. As a prey becomes rarer, it costs more to find it and either the predators begin to die out or they find something else to eat. There is a comforting wisdom in this balance of nature. But it does not even remotely describe the world's fisheries, grazing, and forestry.

Adam Smith considered the subsidies to the Scottish herring industry in his 1776 book *The Wealth of Nations*.[22] He called them "bounties." Herring was an important part of the food of the common people. Subsidized herring might, he hypothesized, "contribute to the relief of a great number of our fellow-subjects, whose circumstances are by no means affluent." He rejected this hypothesis by concluding that "the legislature has been very grossly imposed upon" in paying bounties. Bounties had the effect of encouraging "rash undertakers to adventure in a business they do not understand," knowing that the government would bail them out if they failed.

Two centuries ago Smith understood that subsidies encourage people to enter a business that cannot sustain them. Too many large, winterized trawlers hunted cod off Newfoundland. Once into the fishery, it is hard to get them out. Two months before the cod moratorium, the fishers were preparing their boats for the season with money loaned from local banks and businesses. No fish, no loan payments.

Vagaries of Nature and Acts of God

Some 159 nations have signed, and a further 53 have ratified, the UN Conference on the Law of the Seas III. It requires that they "shall ensure through proper conservation . . . that the maintenance of the living resources in the exclusive economic zone is not endangered by over exploitation."[23] The promise of sustainable fisheries is there.

This promise is not fulfilled because nature is variable, uncertain, unpredictable, and capricious. When fish or forests or grasslands decline, the pressures to continue exploiting them are intense. That is true whether or not the exploiters are subsidized. But as Smith understood, subsidies keep exploiters afloat in bad times.

The agreed aim is to achieve a sustainable harvest, one that will allow the stock to remain constant from year to year—and to keep it as large as possible. A small harvest may mean one of two things. First, there are lots of fish in the ocean, but we are harvesting too few of them. The second is that there are few fish and so they cannot produce enough for us to catch. It follows that somewhere in between there must be a sustainable harvest that is the maximum possible. Exactly the same arguments apply to forests and grazing lands. So how *do* we achieve that maximum sustainable yield?

The vagaries of nature and rare, unforeseen events ("Acts of God") mean that maximum sustainable yields vary from year to year. Variability hurts in several ways. The obvious way is when times are bad. The harvest must be reduced or else the fishery will decline, but fishers still have to live and to pay their loans.

The second is that variability diminishes the credibility of those who argue for caution. No scientist can be sure what the future will bring. If the scientists' estimate of fish that can be caught is too high, the population will decline, which leads to a much-reduced quota the next year. Too low an estimate and fishers will be angry that scientists are preventing them from making more money. Large fishing and fish-processing companies, the small independent fishers, and the politicians all want the highest quotas. Those who lend the fishers money or sell them groceries are hardly likely to disagree. Scientists who urge caution with their uncertain estimations of nature's uncertainties confront formidable adversaries.

Less obvious, but perhaps even more pernicious, is when times are good, for they ratchet up expectations. Such uncertainty appears to have played a role in the cod's decline. In 1982 Canada's Department of Fisheries and Oceans was predicting a quota of 400,000 tons for 1987 and a long-term harvest of 550,000 tons—numbers roughly twice the historical average. Later, these estimates had to be revised sharply downward.[24] The good times run for several years; but then so, too, do the bad. Just like the Nile, understood by Joseph, just like the grasslands that destroyed would-be settlers, just like everything else on Earth.

Becoming Wise

Fisheries have a long history and a clear market value and thus provide compelling case studies of how we destroy what we value. They offer individual and countable cases to emphasize the repetition of our folly and lessons on how to avoid future mistakes.

Lesson Number One

Consider this book's key figures: Human beings use 40 percent of annual terrestrial plant growth, 60 percent of accessible freshwater runoff, and 35 percent of the oceans' continental shelf productivity. These are large numbers, especially since our population will likely double by midcentury. Yet it is not the numbers per se that should alarm us. It is what we have to do to get them, year after year, in good years and in bad.

We harvest fisheries that require only 2 percent of the open oceans' plant productivity to support them. Even at this modest level of exploitation we are still overharvesting the stocks. In more productive seas, we harvest fisheries that require 35 percent of the local plant production to support them. This level has caused widespread declines. The parallels to land are obvious. Grazing lands can be trashed even when our domestic livestock take only 10 percent of the annual plant growth. The first lesson is that these percentages are meaningless unless we understand their consequences. Those consequences can be severe declines in natural resources even when we seem to be getting very little from nature.

Lesson Number Two

The major mechanisms that lead to these declines are not new ideas, as I hope I have made plain. Indeed, they are the excellent case histories: *Bad Land* for grazing lands; *Cadillac Desert* for drylands and irrigation; *Song for the Blue Ocean* for fisheries; two movies, *The Burning Season* about rainforests and *The Perfect Storm* about fisheries. These and many other examples detail the economic mechanisms and the political environment in which they play out. Their real villains so damnable that you hope Hell has a special place for them, are drawn more finely than I could hope to do.

What is not apparent in these case histories is the scale of the environmental impacts they encompass. Take perverse subsidies, for example. They encouraged damage to grazing lands (free land), to rivers (free, or cheap, water for irrigation), to fish stocks (subsidies), and to rainforests (tax breaks for cattle herders and loggers). Their consequences are writ across the 40 million km^2 of drylands, 2 million km^2 of irrigated lands, 300 million km^2 of oceans, and 10 million km^2 of rainforests.

On a small scale we may have considerable enthusiasm for this chapter's protagonist. We may applaud his business acumen, wish we owned stock in his company, and envy his overnight stays with the President or his knighthood. But when two-thirds of the world's major fish stocks are depleted, shouldn't we temper that enthusiasm? The second lesson is that we can no longer afford to waste huge sums of money on actions that damage, perhaps permanently, our global environment.

Lesson Number Three

Scientists can easily invent methods of harvesting fish—or forests or grazing lands—that are benign in the face of nature's caprice. Finding such methods that are also economically acceptable and politically practical is a much greater challenge.

Politicians succeed by pleasing everyone all of the time. If fishers want a catch of 200,000 tons and fishery's managers want only 100,000 tons, the political answer is a compromise somewhere in-between, and probably closer to the higher value. (The same compromise applies to how much forest to log, how intensively to graze land, or how much carbon dioxide to put into the atmosphere.) Compromise is generally a fine thing, but sometimes has as much relevance to a scientific problem as a senator who announces that the sun usually rises in the west because that is what most of his constituents think. Ignoring scientific realities does not matter until someone's livelihood is on the line. If human impacts were not as large, scientific realities would be easy to ignore. John Crosbie was in the wrong place at the wrong time when livelihoods were on the line. His counterparts worldwide are increasingly going to find themselves in the same situation.

Blaming politicians is not my third lesson. Rather, it is that the solution to these problems can be achieved by not only understanding the ecological limitations of natural resources and how we exploit them but also how economic and political factors constrain our actions. That means effectively communicating the science to the politicians—"I'm sorry, senator, but the sun always rises in the east"—and to the public whose views influence the politicians.

What would Huxley have thought about all this? Before you draw the inevitable conclusion about Huxley's naïveté, consider how clouded our crystal balls are in the twenty-first century. Much of what we predict about the future will be in error a century from now. Will we consistently underestimate the damage we may cause, basing it only on extrapolations of current technology?

Despite these chilling statistics and all the shortsighted mechanisms that encourage degradation of natural resources, there is some hope. We can pull back and begin to understand the interacting social, economic, and ecological reasons for our dismal management of natural resources. We have done nothing yet that we cannot correct. The fisheries can recover; forests can regenerate, and so can dryland soils and vegetation. Unfortunately, other human impacts are irreversible. When we deplete the variety of life on Earth, it is gone forever. It is that variety, and how we diminish it, that I will now address.

Part Three

THE VARIETY OF LIFE

The setting is High Table at Balliol College, Oxford, early in the last century. (*High* refers to the table's elevation—it is where the faculty sits for meals—and most certainly not to the quality of the food.) The protagonists are the cleric and Master of the college, Benjamin Jowett, and British biologist J. B. S. Haldane, an acknowledged atheist and enthusiastic communicator of science to the public. The setup is Jowett's question: "What could one conclude as to the nature of the Creator from a study of His creation?" Haldane's reply, "an inordinate fondness for beetles," is the punch line. Since there are more kinds of insects than any other species and almost half of all insects are beetles, Haldane's quip was apt.

University of York ecologist Mark Williamson quickly pointed out that the Haldane story, as told by Robert May in *Nature* in 1989, is a complete fabrication.[1] The time and place are impossible. Haldane was less than 1 year old when Jowett died. Mark told me that Haldane probably said "a special preference" not "fondness," that he was surrounded by fellow scientists in London, not Oxford, and that he was considering the likelihood of finding life (and thus beetles) on other planets. May's reply to this incontrovertible refutation of his facts was worthy of Haldane himself. "Mundane constraints of time and space do not apply to stories about Oxford."[2]

Chapter 11 explains how we know such facts as "There are more kinds of beetles than any other species." It tries to answer the question that children ask: "How many kinds of animals and plants are there?"

Despite his inordinate fondness for impossible historical anecdotes, Australian-born ecologist Robert May has played a major role in answering that question. A child's question can be simple, but that doesn't mean we can answer it easily. There are certainly many more kinds of animals than plants. Although we have not counted them as carefully as the more familiar birds and mammals, there is an extraordinary number of kinds of beetles and other insects. Most creatures are small; few are great. We do not have the names for most of the small ones.

Chapters 12 and 13 focus on "constraints of time and space." How long does life last? And where is the greatest variety of life? Answers to both questions are necessary steps toward understanding how fast humanity causes that variety to shrink.

Despite all the uncertainties about how much variety the planet holds, we know that more than 95 percent of that variety is now extinct. Paradoxically, extinction is a normal part of life. Does this mean that the extinctions we see today are to be expected? Are endangered animals and plants just those whose time is up? Or do the impacts on forests, grasslands, rivers, and oceans contribute to these extinctions? How can we tell?

How long should each kind of animal or plant live from when it first appeared in the rocks to when it disappears forever? The answer is a few million years; to witness this, visit that famous slice of time, the Grand Canyon of the Colorado River, and climb a meter or so. Chapter 12 will show that humanity has shortened animal and plant lifetimes by about 1000 times—down to the equivalent of a few millimeters of the Grand Canyon.

Chapter 13 finds that the greatest variety of animals and plants is in the tropics, particularly in the moist forests. There is a precise, quantitative connection between the amount of forest and how much variety it contains. The forest is shrinking fast, and so too is the variety of life. We need a term for all this variety of life, for all the kinds of plants and animals, for the variety within each kind, and for the diverse places—forests, grasslands, tundras, and so on—in which they live. It needs to be succinct and preferably catchy. In 1980, while pondering this problem for a report he had to write for a U.S. government agency,

Elliott Norse plumped for *biological diversity* or *biodiversity* for short.
The term was an instant and persistent success. These final chapters are
about biodiversity.

Today, the real descendents of the entirely mythical conflict
between Haldane and Jowett have carried the battle to the trunks of
their cars. Some carry small metal icons of fish. The Greek word for
fish, *Ikthus*, is an acronym widely employed by early Christians in
Rome. (*Iesous Christos cotheou hyios soter:* Jesus Christ, Son of God,
Savior.) Some icons have "Jesus" written inside to aid those unfamiliar
with ancient Greek. On other cars, a similar icon has evolved
amphibian's legs and has "Darwin" inside. Some have a fish eating
an amphibian.

Is this funny? Perhaps, but arguing about the mechanism of the
birth of the variety of life spectacularly misses the point. The mechanism of death is the issue. The human impacts on "stuff" have all been
reversible. We could use plant production, land, water, and forests
more efficiently. We could stop repeating the follies of overgrazing and
overfishing, and so on. "Stuff" can recover. Extinctions are irreversible.
Jurassic Park is merely a fantasy. The salient feature of the twenty-first
century may well be how much of that variety humanity destroys, for
ever and ever, with or without an "amen." This should be a common
concern to the descendents of Jowett and Haldane.

11

An Inordinate Fondness for Beetles

The large conference room in Osaka, Japan, was packed long before I arrived, so I must stand outside, behind people who are crowded into the doorway. Craning my neck, I catch only occasional glimpses of the speaker. A single light shines down from high in the ceiling and gives him a saintly glow in the darkened room. I can barely hear the talk, but there are murmurs and laughter. The latter may be at my expense, generated by a caustic remark on the talk I will be giving later on an insignificant fraction of life's variety, the 10,000 kinds of birds.

Spoken numbers are passed to the back of the room: "Did he say a million kinds of insects?" "Five million?" The audience falls silent for a while and then gasps. One of the outside crowd turns to me and asks, "One hundred million?" The speaker's halo, the attentive crowd, the whispered and often mangled information passed to its fringes recalls Monty Python's *Life of Brian* and the scene where cheese makers are assumed blessed. Sir Robert May, former physicist, known for his brilliant insights into ecological theory, his dubious historical anecdotes, and (until recently) his position as the British Government's Chief Science Advisor, is doing the numbers.

Those who venture into the field and forest, whether to watch birds or collect wasps or fungi, are connoisseurs; they are collectors of the exquisite and subtle differences carefully listed in their catalogs. Rumors of May's ever having been in field or forest, where biodiversity is commonly thought to reside, have proved difficult to confirm. He is that rarity, an ecologist without a life list of something, someone without a personal catalog. So it is dispassionately and with a unique virtuosity that he juggles the large numbers—some certain, some uncertain, some quite preposterous—that constitute the counts of described species.

185

The Making of Species

Species are the different kinds of animals and plants. Each species is distinct; its individuals can breed with other members of their own species but not with other species. Counting species is not the only way to estimate the variety of life, but it is the simplest and by far the most common.

So what makes a species? How many are there? And can we go to a central catalog with all the species listed and look for when and where each species became extinct and so measure humanity's impact?

Species enter the scientific domain when they are collected and described. The latter is a technical term. A *description* is a scientific paper that describes the species and how it differs from similar species, where the new species was found, the museum where the specimen resides, and other relevant information. This is the trade of taxonomists, a group that has become scarce in recent years.

David Etnier is my closest taxonomist friend and he specializes in fishes.[1] His laboratory is filled with huge numbers of glass jars, each containing small fish pickled in formalin and bearing labels. In life, they are the brightly colored darters of the small rivers that flow from the southern Appalachians. Their end came when David and his students waded into the water and caught them with nets.

In the Little Tennessee River one day, David found one that he did not recognize. Comparisons to all the other specimens, and their scientific descriptions, explained why. It is a new species. David published the description and gave it two names. The snail darter is the "common name" and *Percina tanasi* is the scientific name. That two-part scientific name is an essential part of the description. It is unique. The first part, *Percina*, however, need not be, for this is the genus to which the species belongs. Genera may contain only one species, but some contain thousands. The genus contains the species' closest evolutionary relatives, but the description does not stop there. It also contains the family to which the genus belongs, the order to which the family belongs, and the class to which the order belongs.

The broadest divisions of organisms are familiar to everyone. Birds, mammals, reptiles, and amphibians are classes, for example. Many familiar plant names from anemone through geranium and magnolia to zinnia are the names of genera. Other familiar names match families: hawks, rabbits, lilies.

Not to put too fine a point on it, how does David know his species is a new one? Just how certain is David that his fish is not the same as one in someone else's collection of glass jars or described in a paper that he has overlooked?

Checking to see whether your specimen is different from someone else's may require visits to the many distant museums that house all the previously described species. The problem of *synonymy*—namely, giving the same species different scientific names—is widespread. A taxonomist may be so overwhelmed by her task that she may mistakenly give another name to a specimen of a species she described in earlier years. Typically, a fifth of all the insect names are synonymous. For some groups, such as moths, nearly half are synonymous.

Even if David is sure there isn't a named specimen like the one he has, could it be that it is just the male or female of some species already described from another specimen of the other sex? Could it be the young of a species previously known from an adult? Knowing how to sex individuals helps, but it is no easy task for some groups. It helps to collect enough individuals to capture a full range of ages and sexes. Some species vary in size, shape, and color over their geographic ranges, so a specimen from the north may look different from a specimen from the south.

From the perspective of our trying to count the number of species, it matters how many specimens of each species are available. We know most species from one or, at most, a few specimens. Nigel Stork, an ecologist at James Cook University, Australia, had wanted to know just how many. He told me that he had opened a large sample of specimen drawers at the Natural History Museum in London and looked at the beetles within. He found that about two-thirds were known from just one place and most of them from only two or three specimens.[2]

Having so few specimens of each species causes one set of problems. In my rash youth I made a brash attempt at humor at a faculty party:

Question: What does a taxonomist call a three-legged mouse?

Answer: A new genus.

The group included the university's vice president. He was a taxonomist of small mammals with a reputation for describing new species—and even new genera—on the basis of one or a few individuals. While others in the group laughed, he did not think my joke was funny, and a career-threatening silence descended.

Only experience gives taxonomists intuition about what to expect. We know that a mouse with only three legs has accidentally lost one because we have seen a lot of other kinds of mice that all have four legs. Experience is not won easily. It takes time and effort to work out which specimens are male, female, young or old, or from different parts of their geographic range.

Those who describe fossil species sometimes don't even know the physical orientation of the living animal—which end was up. The vice president knew from a lifetime's experience how difficult it is to collect enough specimens to satisfy his critics. He did not need some punk assistant professor reminding him.

Not all taxonomists make the same choices. Some are so impressed with subtle differences in shape, size, and color that they describe new species every time they encounter such variation. One botanist in Hawai'i is particularly notorious, claiming dozens of species that everyone thinks are individual variants of the same species. Human beings come in a variety of sizes and shapes and have different eye, hair, and skin colors; they are not all different species. Once again, it takes effort to show that human variation is small compared to the differences among humans, gorillas, and chimpanzees.

Taxonomists have a second set of problems when many specimens are available. These are typically large species (birds, mammals, lizards) or brightly colored ones (butterflies, marine snails, flowering plants). Similar species can be geographically discontinuous. Different islands in Hawai'i or in the Caribbean, for example, have similar species of birds. Different river basins provide examples of David's new fish. Are these really different species, or merely local forms of the same one?

When described in the late nineteenth century, the fashion was to name each geographic population as a different species. The name was often that of the financial backer of the expedition: Five bird species are named for Lord Derby, five for Rothschilds.[3] The more names there were, the more patrons could be flattered.

By the mid-twentieth century, taxonomists began to deplore this proliferation of names and followed the lead of the influential Harvard evolutionary biologist Ernst Mayr.[4] By his definition, all members of one species had at least the potential to interbreed. If only the amakihi I saw slurping mamane nectar on Maui could meet the ones on the neighboring island of O'ahu, it would be love at first sight. They never do meet, so the decision is completely arbitrary. The new fashion was for geographic isolates to be considered forms of one species. Their names were "synonymized."

Lumpers and Splitters

Those who group forms into one species are called *lumpers;* their rivals, who consider each isolate to be a different species, are called *splitters*. If this schism sounds like something from a satire by Jonathan Swift, then you're getting the right idea. Lumpers receive hate mail from birders who, with fewer species on their American life lists, are unceremoniously thrown out of the prestigious 700 Species Club. Birders should take heart. Lumpers, who have been ascendant for so long, are being challenged as never before by DNA evidence.

It is a straightforward matter to observe the differences in the DNA of the amakihis on Maui from those on O'ahu. We can even date when the two populations were separated, because differences in the DNA accumulate over time at a steady rate. There is a molecular clock: the longer the isolation, the greater the difference in the DNA.

Working backward, if we know how different the DNA is, we can calculate how long it has been since the birds on O'ahu bred with those on Maui. The answer is on the order of millions of years. The separation probably came about when a small group of birds was blown from one island to another, thereby founding a new population and forever isolating the two groups.

Arguing that populations have the potential to interbreed is fine if you think the differences are small and, by implication, recent. Just because you haven't visited your second cousins for a few years does not mean that they are not part of the family. Separations of a million years or more is another matter. There is not much potential in that length of isolation. Surely if there were potential it should have been realized from time to time and the populations would have swapped a few genes. Perhaps we should consider geographically separate populations that have been completely isolated for long times to be separate species. Many of this century's taxonomists agree.

The consequences of all this are illustrated by Bruce Patterson, at the Field Museum in Chicago. He has been looking at names of mammals.[5] In the last 20 years, 60 new species were described as new to science. Sixty-two names were banished as synonyms. Yet nearly three times that number were resurrected as new species from the scrap heap of earlier synonymy. For a typical member of the largest group, a nineteenth-century taxonomist had said, "This is a new species." A mid-twentieth-century taxonomist had proclaimed, "No, it isn't" and now a late twentieth-century taxonomist was asserting, "Yes, it is."

Counting Species

What chaos! With no agreement on names, how can we count them? Surely in such confusing times a visionary must come forward! It helps if such a person is not deeply mired in personal disputes of this split or that lump. Enter Sir Robert May.[6]

First of all, May is critical of misleading precision. One species count was "1,392,485 described species." The problems of synonymy and arguments between splitters and lumpers make it doubtful that the correct total is accurate to within 10 percent. Against this background of uncertainty he presents the approximate totals:

> There are about 70,000 species of fungi, 270,000 plants, and 40,000 algae.
>
> There are 15,000 species of nematodes (tiny worms, some of which parasitize plants and animals—including us), and 70,000 mollusks (snails, clams, squid, etc.).
>
> Arthropods win hands down. They include 75,000 species of arachnids (spiders, scorpions, mites), 40,000 crustaceans (shrimp, wood lice, crabs), and a whopping 720,000 insects. Of these, 400,000 are beetles, 150,000 are moths and butterflies, 130,000 are wasps and ants, and 120,000 are flies.
>
> There are only 10,000 species of birds and 5000 species of mammals. These are the species that we know the best, but birds and mammals—indeed, all the vertebrates combined—are only a small fraction of named species, perhaps totaling 45,000 species.

May estimates a total of about 1.5 million species. Other scientists derive totals between 1.4 and 1.75 million species. These are big numbers, but they do not surprise his audience. Nor do his statements that taxonomists make outright errors. And, yes, we understand their differences of opinion over whether to split or lump geographically separated populations. But even if taxonomists had their houses in order and errant members of the profession were placed under house arrest, and they received bipartisan support for an agreement on splitting and lumping, they still would not give us the answer to how many species there are.

What surprises May's audience the most are his estimates of how many species there are for which we have no specimens at all. Even the smallest are in the range of several million species, while the largest reach 100 million. He pinpoints the exact problem. "Taxonomic effort reflects passion, not ecological importance." Using Britain as an example, he quips, "Fascination with the furries and featheries goes deep. The Royal Society for the Protection of Birds has almost 1 million members; the analogous society for plants has around 10,000. There is no corresponding society to express affection for nematodes."

The differing affections are reflected in how many new species taxono-mists describe each year. For birds, centuries of passionate pursuit mean that only a few new species are found each year. The consequences are pre-dictable. Bird names are sold off by auction. The asking price for a recently discovered bird was $100,000. (Rabbits also find ready buyers: Hugh Hefner of *Playboy* fame has an eponymous bunny from Florida.) The cost includes the services of the taxonomist to write up the description and name it appropriately. The cost is tax deductible since the money is sent to conservation organizations in the country where the species was found. Immortality never came so cheaply, for scientific names are a permanent memorial. (Perhaps Lloyds should offer insurance against becoming a syn-onym.)

Rosie Gillespie's Spiders

Apart from birds, there is nothing approaching a complete taxonomic cata-log for any other species. If only Rosie Gillespie at the University of Hawai'i could charge for all the new spiders she is describing! We sit outside her apartment near the university enjoying the tropical sunset. "What did it feel like to find all these species?" I ask enviously. "Pretty intoxicating, *actually!*" Rosie, another transplanted Briton, had come to Hawai'i to look at the for-aging behavior of spiders and became intrigued by the number of *Tetrag-natha* in the forest. Spiders in the genus *Tetragnatha*, meaning "four jaws," had been the subject of her dissertation.

"I was convinced that I had a good feel for them . . . and they did not live in the forest!" *Tetragnatha* worldwide are similar in shape; they are almost all brown and elongate. They have similar life styles, such as dwelling over water, and they have similar behavior, such as building flimsy orb webs. . . . But on Maui, the *Tetragnatha* were living in the heart of the forest."

Soon after she started her work, fellow spider enthusiast Vince Roth came to visit. That night, stepping outside in their raincoats and Wellington boots they started collecting. Rosie went on with the story:

"The spiders are strictly nocturnal. It is really hard to see them during the daytime because they conceal themselves in crevices on tree trunks, on the ground . . . and goodness knows where else. So the only effective way to look for them is by flashlight at night, when they become active, running around or building webs all over the place."

Rosie showed Vince the forest and the fascinating forest-dwelling *Tetragnatha*. He started looking at the other spiders in the forest. . . . "Were they *Tetragnatha* as well?" he wondered. Rosie and Vince took their catch back to her house in Pukalani and looked down a microscope. They were indeed. "And the amazing thing was that not one of these large, conspicuous spiders that dominated the open forest at night had been described."

With more research, Rosie came to recognize that almost all of the spiders found in the open in all the Hawaiian forests—whether wet, dry, or intermediate—and on all the largest islands—Kaua'i, O'ahu, Moloka'i, Lana'i, Maui, and Hawai'i—were *Tetragnatha*. Unlike their mainland counterparts, they could be long and thin, squat, round, or bumpy. They could be green, black, red, or pearly white, and they ranged in adult size from a couple of millimeters to a couple of centimeters. And they had peculiar habits.

Some had modifications of the jaws, apparently to allow them to specialize in specific kinds of prey. One species on the top of the Ko'olau mountains, one of O'ahu's ranges, has jaws without the usual teeth. Instead it has serrations along the margins for capturing amphipods. (*Amphipods* are crustaceans; sandhoppers on beaches are perhaps the most familiar.) Most built webs, but several groups of species had abandoned normal web building and were mobile predators that chased their prey. Yet another species used elongated claws to impale prey, often striking them directly from the air.

Rosie knew that the taxonomists of the last century had counted a grand total of nine species. Before the evening with Vince was over, they knew there were many more. By 1994 Rosie had described 19 new species and had found more than 60 new ones.[7]

Counting the Uncounted

Stories like Rosie's are played out throughout the world as taxonomists collect specimens. Her experience is of a familiar group—there are societies that express affection for spiders—and in a well-known place—the United States. She still found lots of new species. Elsewhere, perhaps a large majority of taxonomically unfashionable species might not yet be known to science. How can we tell? How can we estimate what fraction of species is unknown?

One solution is just to ask experienced taxonomists to guess how large a task they have ahead of them. Or, we can invent quantitative methods to add the thinnest veneer of statistical respectability to these wild guesses.

British mycologist David Hawksworth pondered about the number of fungi there might be.[8] His thoughts ran like this. Britain has about six times more species of fungi than of flowering plants. Britain's fungi and flowering plants are well known, a consequence of a long tradition of natural history, so this 6-to-1 ratio seems reasonably secure. Applying this ratio to the world total of 270,000 flowering plant species, Hawksworth estimated 1.5 million species of fungi. Only about 70,000 species of fungi currently have names.

May is quick to cross-check Hawksworth's wildly uncertain number. If there really are 1.5 million fungi and only 70,000 have names, then only 5 percent of the fungi in the world should have names. So collect fungi from poorly explored places and check the fraction of them with names: It should be 5 percent. In support of his estimate, Hawksworth has written of his experiences in Kenya, a country likely to have poorly known fungi. In his sample of fungi, about 30 percent had names. That is not very many, but it is too many to support his claim. If as many as 30 percent of the world's fungi have names, the corresponding estimate for the world total would drop to a more modest 0.25 million species.

May does not flatter Hawksworth, considering his estimate to be "very discordant with the facts." May also has choice words about estimates of 10 million or more species of mollusks, crustaceans, and worms in the mud of the world's oceans. Nor does he like the suggestions of tens of millions of insect species. It is the rejection of the highest totals that brings his assessment of the estimates and calculations to a simple conclusion.

There are about 4 million species of insects, a "best guess" of 7 million species in total and a plausible range of "5 to 15 million species." The best guess consists of several parts that can be compared to the numbers of described species.

There could be about 500,000 species of fungi, 320,000 species of plants, and 300,000 species of algae. Compare these numbers with the numbers that have already been described above. If these numbers are right, it would mean that most plants are described, but most fungi and algae are not.

There could be 500,000 species of nematode worms and 120,000 species of mollusks. These are not widely studied, and this number is much higher than those that are described.

Arthropods still win hands down. There could be 500,000 species of arachnids. (The large numbers of arachnids mean that Rosie Gillespie's experience will be repeated again and again for centuries to come.) In addition, there could be 150,000 species of crustaceans, and a whopping 4 million species of insects. Of the insects, most would be beetles.

The vertebrate catalog is nearly complete; there could be a total of about 50,000 species.

May is not the only person in this numbers game. Nigel Stork has gone over the same ground. His take on these numbers is a "best estimate" of between 12 and 15 million species.[9]

Taxonomy's Final Frontier

The greatest controversies in the total species count are not from fungi or the ocean mud. They come from taxonomy's final frontier—the canopy of the rainforest. Rainforests used to be remote, but getting to one these days is no longer that difficult. Advertisements for tours of the Amazon or Costa Rica are common enough, allowing the visitors to boat past a rainforest or walk along trails through one. The canopy above you is quite another matter: How do you get up there?

The forest floor is dark, its still air rich with the smells of decaying wood. Ahead is a tower. I can see only its first few stages. Each has two sides, a narrow metal ladder from one stage to the next and a horizontal plywood board 1 m wide and 3 m long. Climb the ladder, walk along the board, climb the next ladder. The boards have rotted in the damp heat, so I must waddle, placing my feet carefully at the sides where the boards are supported by the metal scaffolding. The first few stages are easy. Then the distance to the ground becomes obvious, and my pulse starts accelerating from the effort; the sweat on my body doesn't evaporate to cool in the humid air. I've been here before: This is the worst part. Some of the students who have come with me change their mind and climb down.

A few more stages and I am in the foliage. The ground is invisible, no longer a threat, and I relax. The humidity drops, the air freshens, bright green life takes over from brown decay. The foliage is so varied that no two trees look alike. Their leaves are different sizes, shapes, and colors. After the darkness of the forest the sunlight is brilliant. The early morning mist still hangs in patches above the canopy, blowing around in the breezes.

A breeze catches the tower and cools me. I turn around. How did the low wave of dark clouds appear so quickly? A few spots of rain are the forerunner of a thunderstorm, and I am standing on an all-metal lightning rod 35 m above the ground. We climb down as quickly as we dare, and in the last few stages we are cold, soaking wet, and in near-complete darkness. "We should come back this afternoon." The rain is so heavy that I must shout against the storm. No one disagrees.

When we returned in the late afternoon sunshine, it was to watch brilliant green parrots and toucans, scarlet macaws, iridescent blue honeyeaters, and multicolored tanagers. Rainforest canopies are hard to visit but worth the effort. This one spot in the forest canopy has more bird species than I could see in all of the eastern United States. All we've seen so far are the large, obvious species. I can see, but cannot touch, the trees around me. How many insect species are up there is anyone's guess.

During the day I want to teach a lesson about trees. "Walk through the forest and collect leaves from the first 20 trees you encounter." Off go the students, returning before lunch with their samples to sort them into different piles, one for each species. This is an exercise I have used in other places. In Britain, students may find a dozen species; most they will know by name. In the richer forests of Tennessee the total will be three times that, and everyone will know the broad classes of oaks and hickories, if not the particular species. By now, lunch is ready and we step over the several dozen piles of leaves on the concrete floor under the tables. Balancing a plate of rice and beans in one hand, a leaf and a fork in the other, students walk among the piles looking for a match. When they are certain there isn't one, they start a new pile.

They are still sorting leaves after a dinner of yet more rice and beans. By 9 o'clock in the evening, we finish the count: 200 kinds of leaves. Almost all will be different species, and the students know the names of almost none of them. Naming tropical trees is no easy task, and we call over one of the forest reserve's professional botanists to look at our efforts. The botanical team has nearly completed a survey of this 100-km^2 reserve, so I ask about their total. "Two thousand species of trees" is the reply. "Some of your piles are more than one species. You can separate these species only by their flowers, up in the canopy."

Off to the canopy goes one of the botanists, with a companion whose job is to string the ropes into the trees. Having worried about climbing the tower that day, I am completely outclassed. This young Brazilian will climb into the canopy of the Amazon rainforest by rope, alone, and at night. That is when the species she studies is in flower.

Those efforts of climbing by ropes strung into the canopy have paid off. Identifying the reserve's 2000 trees is now possible; there is a written guide. Nonetheless, it is a daunting task for an experienced tropical botanist. There are more families of trees than Britain has species.

Counting and identifying the insects on the leaves of the trees are likely to be hundreds of times more difficult. That counting and identification cause the arguments about how many species there are. The canopies of each one of the 2000 species of trees will have insect species that live on no other tree species.

Ecologist Terry Erwin of the Smithsonian Institution leads the field of high estimates of insect species.[10] He drapes large plastic tents over rainforest trees and does "canopy knockdowns." He fills the tent with an insecticide that kills the insects on the tree; they fall to the ground and are collected. In theory it sounds so simple.

Just one species of tropical tree held 1200 species of canopy-dwelling beetles. The crucial statistic is that 163 species were found only on that tree species. This fraction of about 14 percent is called the *host specificity*. Other beetles in the sample were not specific to one host; they lived on other tree species as well.

About 50,000 species of tree are in the tropics, so Erwin multiplied 163 times this number, to get about 8 million species of canopy beetles alone. About 40 percent of described insects are beetles, so the total number of canopy insect species could be 20 million. Adding half that number for insect and spider species on the ground gives a grand total of 30 million species.

Opinions range about the relative numbers of beetles and of species in the canopy and on the ground. The most detailed studies show that four times as many species are on the ground as in the canopy. Using that figure gives the 100 million species estimate.

Erwin's method is a stepwise multiplication of a string of large and very uncertain numbers. The most uncertain one of all is the 14 percent of beetles specific to one kind of tree. Reduce that number and the estimates come tumbling down.

May asks the same question he asked about fungi. What fraction of the species found in canopy samples are new to science? If there really were 8 million kinds of beetles in the canopy, and only 400,000 had names, then only 5 percent of Erwin's beetles should have names. The number appears to be nearer 20 percent. Erwin's estimate of 30 million is too high, May concludes.

Once again, May's careful deliberations and meticulous cross-checks of the numbers generate a smaller number than the taxonomists', and I wondered why. There is a hint of the "drunk under the lamppost" story. The drunk (taxonomists in the analogy) looks for his keys (species) under the light because that is where he can see better (species and places taxonomists know well) and not where he lost them (species and places taxonomists don't know well).

Roger Kitching works in the Queensland rainforests near his home in Brisbane. He is an ecologist and so must look at species where they are, not where the light of taxonomy illuminates the darkness. I wanted to know his views about the "speculators": May, Stork, Erwin, and the others. Roger supports Stork's estimate, but not Erwin's: "I feel it's much better than May's 7 million odd, but I am still left with substantial uneasiness. Let me try to explain why."

Roger tells me that "most field workers, including even Terry Erwin, underestimated and misunderstood the mites entirely, throwing them out as too difficult." In the subtropical rainforests of Lamington, south of Brisbane, the results from canopy knockdowns suggest a total of about 2000 species. Seventy percent of these species are new and include seven new genera. Yet even these are among the better-known groups of mites.

"Much of what I've said about mites can be reiterated for flies." Roger goes on. "They are generally sidelined in favor of the better known and 'easier' beetles and moths." Roger has just returned from a seminar on a group of fruit flies. Detailed studies were showing that many of the standard species needed to be split because they were actually complexes of many species, each requiring minute study to resolve them. That resolution would multiply earlier totals of described species sometimes by tenfold. "Many of the numerically most dominant fly families in rainforests are among the globally least well-known, and more doubt is sown about any estimate."

Roger is telling me that the Creator's alleged fondness for beetles might not be in jeopardy, but mites and flies might be more appreciated than taxonomists had perceived. "Working out the numerical relations for beetles and then basing estimates on them simply will not work."

What I really want to know are his views on Erwin's extrapolation from the number of species specific to each tree species. It is a particular interest of Roger's and the linchpin of the highest estimates of species numbers.

"There is little doubt that Terry Erwin's estimates of 14 percent host specificity among the beetles is an overestimate. Values of 4 to 6 percent seem more likely. But this infelicity is overwhelmed by the fact that Terry spoke only of trees in his famous stepwise multiplication. He ignored herbs, vines, epiphytes, lichens, ferns, mosses, fungi—what an omission!—as possible hosts for insects, mites, and spiders."

The epiphytes—plants including orchids, bromeliads, and ferns that grow on trees—had been a common research interest of ours a decade earlier. Of the mites and insects we had discovered living in the water-filled crevices of some of these species, several now bore the scientific name *kitchingi*.

Roger does not know how host specific the animals might be on these other plants and fungi, but one thing is certain: They comprise more species as potential hosts for insects than do the trees. Only a fifth of plant species are trees and even May conceded that there might be more fungi than plants. Roger's final point is that the data that ecologists have is usually gathered from only a few places.

"Rainforest field workers are continually in wonder not just at the richness of the forest they work in, but at how varied is the forest across even the shortest distances. I can make sweeping statements about the canopy I have studied adjacent to O'Reilly's guesthouse, but I doubt if they have much relevance to the forests a little way down the Picnic Rock track, nor to the tree-fern and palm forests farther along. Certainly, they tell us nothing about the beech forests 5 km along the border track. And I wonder what they tell me about apparently identical forest just 3 km down the road."

Erwin's beetles were from one small area. Given the difficulties of getting to the rainforest canopy, how could it be otherwise? The fractions of described species too high for large global totals would all be from accessible locales. The problem with counting species in only one place was whether or not a forest with the same tree species would have the same species of insects.

Towers do not get you close enough to different trees and their epiphytes. They are hard to put up and hard to move. Ropes are rather too exciting. So the solution is a canopy crane. It could deliver a researcher to anywhere within a certain circumference and height; hence it could be moved, though not easily. If the same tree species had different sets of insects, every one specific only to it as well as to a restricted locale, then the total counts of species could be huge. Roger has a crane on order.

The Big Book of Species

There is no catalog, either hardbound or in the computer, of the 1.5 million species for which we have scientific descriptions. Taxonomists are giving the idea serious thought, however. Even if this catalog were to be finished soon, it would only be a thin tome compared to the complete enumeration of life on Earth. If May's predicted total of 7 million species is right, the task of completing the catalog—describing new species—would keep taxonomists busy for 600 years at their current rate of progress. But we don't even know how large the catalog should be.

It seems that when asked, "How many species are there?" taxonomists always guess higher numbers than May says their data will confirm. Samples of beetles and fungi in remote places contain too large a fraction of described species for there to be tens of millions of species. Yet taxonomists sense the uncertainties, for they collect and identify only the groups they know, in the places they expect them. What could be hiding in the taxonomic darkness is impossible to guess. In any case, it is not the taxonomists' problem to count species, and too high a total might discourage them.

Ecologists need species names in order to do their work. Rosie Gillespie wanted them so she could study spider behavior. The insects in the water-filled cavities of rainforest canopy epiphytes that Roger still studies are the vectors of yellow fever, malaria, and likely other diseases ready to emerge from the rainforest when we move closer to its edges. Ecologists inconsiderately look into the taxonomic darkness. They present their glass vials filled with specimens to the resident taxonomist. In response to the obvious reply they retort, "What do you mean you can't identify them? To the follow-up question, "Who can?" there is often the reply, "Nobody."

Roger's instincts about the taxonomists' uncertainties match mine. "I'll go along with 12 million as a conservative estimate, but all the shaky parts of the underlying logic—if they are ever resolved—are likely to raise this estimate. Field workers generally come up with high estimates of global richness." If Kitching is right in his estimate, the task of completing the catalog would take until the end of the third millennium.

Put simply, the Creator's inordinate fondness for beetles and other insects means that we do not have a species count and we are not going to get one anytime soon. Were we to have the complete catalog, then we would have to put a black mark next to some entries with the date that they disappeared forever. We could count the species numbers and count the number we liquidate. With no such catalog in sight, the goal of evaluating humanity's impact on biodiversity requires a different approach.

In many facets of life we make routine judgments on the basis of incomplete catalogs. Estimates of the "cost of living," for instance, do not measure every price of every item traded. Rather, we make those estimates on the basis of small but representative samples of many different prices. So it must be with biodiversity. We can take small samples of species that we know well. Then we must ask, How fast are they dying?

12

. . . Abode His Destined Hour, and Went His Way

I am preparing for a committee hearing of the U.S. Senate on the Endangered Species Act, and I can guess what one of the questions is going to be.[1] The act provides powerful protection against extinction. Someone on the committee is going to say, "Species have always gone extinct, so why are we trying to prevent extinctions?" You cannot deny your questioner.

How many species have come and gone since life began? The total must be huge. The dinosaurs are gone; their resurrection in the movies teaches that they were many and varied. Visits to a nearby quarry, a road cut, or sea cliffs might uncover trilobites or ammonites. The last trilobite disappeared before the first dinosaurs, the last ammonite disappeared at the same time as the last dinosaurs. Both groups succumbed to the same cataclysmic event. Trilobites and ammonites are every collector's favorite, their shapes beautiful and unfamiliar. Nothing like them lives today. Extinction happens.

Deciding whether less familiar fossil groups persist in the present is a more difficult task. Could Victorian writer Sir Arthur Conan Doyle have been right in making tepuis the last refuge of dinosaurs?[2] I have flown over them, shared his sense of wonder, and have lucky friends who have explored them. They have not found dinosaurs.

In a more prosaic vein, can we be sure that the shell found in the quarry doesn't belong to a species living at the bottom of the sea and described in an obscure publication? The answer from many lifetimes of careful work is that most fossil species are long gone. When a "living fossil" is discovered—like the famous fish, the coelacanth—it is big news.

The senator is right: Species have been going extinct for eons. So what is different now? What evidence indicts humanity as the cause of modern extinctions? And how can we draw conclusions when we have names for, at best, only one in five species?

"Senator, death and taxes are unavoidable, but if all of us in this room died in the next hour, or had to pay 90 percent of our income as taxes, something would be very wrong." It's not a matter of absolutes, I tell him— not whether there should be extinctions or not—but a matter of rates— what percentage of species becomes extinct per year.

The Lifetimes of Species

Consider this analogy: Death is a normal part of life. As individuals, we live our three-score-and-ten years, then die. With that average, one individual in 70 dies each year. Put in different units, only one individual in 600,000 should die in an hour. (That is how many hours there are in 70 years.) When I give a seminar to 600 people, there is a 1 in 1000 chance that one member of the audience will die during my hour-long talk. No one ever has. Were it to happen routinely, I would worry about my lecturing style.

Extending the analogy to species' lifetimes suggests a recipe. First, work out how long species live, how long their "destined hour" should be. From that fact, estimate what fraction should die each year. Then compare this to the fraction that does die each year. The comparison measures how much humanity has shortened species' lifetimes, which is the purpose of this chapter.

Paleontologists carefully catalog their specimens, publishing their descriptions in the same way that taxonomists do. They also report the layer in which a species is first found and the one in which it disappears forever. Measure the depth of rock where a species is found; it tells you how long the species lasted.

There are layers within layers within layers. The major layers are the geologic periods: Cambrian, Ordovician, Silurian, Devonian, Carboniferous, Permian, Triassic, Jurassic, and Cretaceous periods follow in a sequence that started more than 600 million years ago, when the first hard-bodied species appeared. Each period averages about 70 million years, but each period is further divided into shorter stages that typically last tens of millions of years. The sequence continues with the Paleocene, Eocene, Oligocene, Miocene, Pliocene, and Pleistocene. They divide up the time at about 10 million years apiece, and they, too, are subdivided.

Geologic periods are defined by the fossils that the rocks contain. The greater frequency of names in the second sequence (the more recent rocks) reflects a more complete fossil record and a finer classification of time. The ends of several periods correspond to major upheavals of life. For example, the first sequence ended 65 million years ago with the abrupt end of dinosaurs and ammonites.

To estimate a species' lifetime we could look at the rocks themselves and measure directly the depth of rocks and therefore estimate the species' lifetimes. This method results in lifetimes that will likely be too short. Look at a quarry or a cliff. Collecting the lowest and highest specimens of a fossil may be easy, but the resulting two questions are not. Where was the fossil before it was here? And where was it afterward? Only if both answers are definitely "nowhere" will the estimated lifetime be accurate. The limited availability of exposed rocks usually precludes this certainty.

Something similar to the fossil may be found somewhere else. It is hard enough to get the taxonomy of modern species right. Fossils are much trickier. Is the fossil in the rocks of Britain the same species as the similar-looking fossil in Texas? Perhaps yes, perhaps no, but surely it is the same genus. Genera, more than species, are the lingua franca of paleontologists, who are more likely to call a fossil by its generic name. It was the plainly named Velociraptor that terrorized the children in *Jurassic Park*. Only the most famous fossil, Tyrannosaurus rex, the species that improbably came to their rescue, merits a full, two-part name in everyday use.

A different approach to calculating lifetimes surveys the published catalogs and counts the subdivisions of periods in which a fossil occurs. Most fossils are found in only one. A typical record might be for a species to occur in the middle Miocene, a subdivision that lasted 5 million years. Did the species last all that time or just part of it? We cannot tell. Future historians may record that you lived in the twentieth and twenty-first centuries, but you won't live 200 years. For the same reason, species' lifetimes deduced from catalogs will often be too long.

In short, rock-based estimates of species' lifetimes will be too short, while catalog-based estimates will be too long—especially those from the earliest geologic periods, which are poorly resolved. The catalog-based estimates give average species' lifetimes that are 5 to 10 million years—pretty much the typical length of a geologic subdivision. Rock-based studies do show shorter species' lifetimes. Studies of Silurian graptolites from the Urals, Cambrian trilobites from the Baltic, and Tertiary corals from the Caribbean all record species' lifetimes as ranging between a few hundred thousand years and a few million.

Fossils of terrestrial vertebrates are much scarcer. The scarcer the species, the less likely that paleontologists will find its first and last members and the shorter its lifetime will appear to be. Most rodent species have lifetimes of a few hundred thousand years; so, too, do horses and their relatives. Species of insectivores, such as shrews, hang on longer, averaging a few million years.

Just how long is the destined span of time that species last, and why are some shorter-lived than others, are fascinating and fundamental questions about life on Earth, but peripheral to my story. We need only a rough estimate. And that estimate, each species' "destined hour," appears to be about 1 million years—perhaps longer, but certainly not much shorter.

A way to cross-check a species' lifetime is to ask how often a species is born. Return to the analogy of human lifetimes, our three-score years and ten. How old are your living brothers and sisters? The answer could be anywhere from "newly born" to "retired." Their average age will be about half a lifetime. For species that live fast and die young—a mouse, for example—half a lifetime will be 6 months. For centuries-old redwood trees, half a lifetime will be measured in centuries. In essence, the age of your siblings is a clue to how long you will live.

Using the Molecular Clock

Species have the equivalent of siblings, which are each species' closest relative in the evolutionary tree. How long ago did what we now call two species—actually called "sister taxa" in the taxonomist's trade—split from their "parent species"—another trade term?

Grind up some tissue, extract the DNA, and see the differences between the two species. The accumulated differences in the DNA act like the gradual ticking of a clock. More differences mean that more time has elapsed since the split. How fast does the clock run? Some splits can be timed from special geologic events such as the isolation of the Pacific from the Caribbean when South and North America joined. There may be enough fossils to date the split.

For mammals, studies of carnivores (such as foxes and lions) and primates (monkeys and humans) broadly support an average origin time for modern species of about 1 million years ago—the same number as from fossils.

So what happens during a baby boom? We will end up with lots of young siblings. This might mean that the population is increasing—analogous to species being generated rapidly. And this may or may not mean that the species will have normal lifetimes. Perhaps we are living in an era that is spawning large numbers of species naturally destined to live much shorter lives than ones in the past.

The huge advantage of dating species' births using a molecular clock is that it works equally well for all species, rare and common. It works for birds, for which we have few fossils but know a lot about modern extinctions.

To some, birds appear to be undergoing just such a baby boom. In North America many pairs of similar bird species live in forests on either side of the central prairies. There is a western bluebird and an eastern bluebird. And there are eastern and western nightjars, swifts, hummingbirds, woodpeckers, flycatchers, and warblers—indeed, so many pairs that a simple, single mechanism was thought to have created them.

In this Creation story, each pair had one parent species ranging across the continent. Then the glacial scalpel of a Late Pleistocene Ice Age severed each species into two, one in the west and the other in the east. When the ice retreated a mere 10,000 years ago, the sister taxa had diverged enough for taxonomists to deem them separate. Instant species: Simply add (frozen) water! All these species in North America are merely a fortuitous baby boom. Critics were quick to argue that current high extinction rates might be a natural pruning of this evolutionary exuberance.

In fact, the suggestion of baby boomer species is wrong. John Klicka and Bob Zink from the University of Minnesota used molecular data to show that for 35 such species pairs, the average split time was 2.45 million years ago.[3] The once plausible scenario of Ice Age baby-boomer species fell victim to molecular technology; the species are too old. Indeed, they were considerably older than the 1-million-year estimate I suggested as a benchmark.

This diversion into the subject of species siblings presents another thread in a still-flimsy fabric that suggests that a species' lifetime should average 1 million years, perhaps more.

Obituaries

Now, if we live 70 years, on average, then—also on average—one in 70 people will die each year. For species that live 1 million years it means that only 1 species in 1 million should go extinct naturally in a given year. So what fraction of species is going extinct each year? By how much have we increased the rate of extinction? For most of the more than 1 million named species, we simply do not know. These species have been collected once, pinned in museum cabinets, and never seen again in the wild, just one step ahead of the even larger number of species never collected at all.

For birds—that tiny fraction of species for which millions of people express so much affection—we do have the answer. When a bird becomes extinct, we do much more than put a mark by its name in a global catalog.

We write its obituary. For birds, we can contrast how fast we write obituaries with how fast we ought to be writing them.

What are the odds that you will see a bird go extinct in your lifetime? The global catalog has about 10,000 species. With the odds set at 1 in 1 million per year, there should be one bird extinction every 100 years (less often if the several million years were the correct benchmark). That is less than a once-in-a-lifetime, once-across-the-entire-planet event, a fact to share with your incredulous grandchildren. "I once read this book by a professor who knew Dr. James Tanner when he retired from teaching at the University of Tennessee. As a young man, Dr. Tanner wrote his doctoral dissertation on the ivory-billed woodpecker, a bird that is now extinct."[4] "Wow, granddad!"

That's it. You are not allowed even one more bird extinction. Just one more, anywhere in the world, would be cause to accuse humanity of the crime of shortening species' lifetimes.

If, as chief inquisitor, I want to make that accusation, where would I look? It is easier to solve a recent crime than an old one. So where will the crimes be freshest? While humans have lived in Africa for 500,000 years or so, and in the Americas for more than 10,000 years, it is only in the last 1000 years that we reached remote islands like Hawai'i. Let's return there.

I am on Barber's Point, west of Honolulu, with Bob Pyle from the nearby Bishop Museum. Unlike most of the volcanic island of O'ahu, Barber's Point is a raised coral reef. It is hard going, and we must pick our way carefully. The surface is potted with deep holes and cracks, much as a submerged reef would be. Centuries before the Polynesians arrived in their outrigger canoes, birds became trapped in these holes and died.

Bob tells me to put my hand into the bottom of a water-filled hole. From the muck, I recover a dozen bones. All are of species no longer found here. Some are of seabirds that now nest only on remote, uninhabited islands elsewhere in the Pacific. Others are of an extinct flightless goose that we know only from O'ahu.

Across the Hawaiian Islands, husband and wife team Storrs Olson and Helen James from the Smithsonian Institution have found bones of 43 species that became extinct soon after the Polynesians arrived.[5] The Polynesians ate the large bird species and introduced pigs and rats to the islands, which made quick work of native bird eggs.

These known losses are only part of the story. Bird bones are fragile and easily destroyed, so the bone finders are likely to have missed many species. How can we count what we don't know?

Michael Moulton, an ornithologist from the University of Florida, and I realized that there is a way around this problem.[6] In addition to the 43 species known only from their bones, Olson and James found bones of more familiar species, ones that survived the Polynesian onslaught and lived long enough to be described by modern taxonomists. But they did not find specimens of *all* the surviving species. They found only about half of them. That is how incomplete their survey was, exhausting though the work had been.

Since they found half of the species that survived the Polynesian colonization, they probably found only half of the species that succumbed. In addition to the 43 now-extinct species known only from their bones, there is the other half, another 43 or so extinct species waiting to be found.

Adding the known fossils to the ones waiting to be found, Michael and I repeated our calculations for other Pacific Islands. As the Polynesians colonized the Pacific from New Zealand, north to Hawai'i and east to Easter Island, they exterminated 500 to 1000 species of birds.

At a scientific meeting in San Francisco the next spring, I met David Steadman, another ornithologist at the University of Florida, who had dug for bones in this area. I told him of our calculations. "You're wrong," he declared flatly. I thought at first that he was merely irritable because he didn't like my telling him that he had missed so many bones. Crawling through tight caves into muddy or dusty holes and digging up bones is no more fun than shivering in the rain atop a rainforest canopy. But that was not David's point: "There could be 2000 species of rails alone missing from the Pacific."[7]

Steadman should know. He and his colleagues have searched dozens of the 800 or so islands in the Pacific, from tiny atolls like Lisianski to large volcanic islands like Hawai'i. Each has yielded at least one species of rail, either living or known from bones, and the larger volcanic islands have yielded several species. From these observations Steadman argued that every one of the islands would have had its own species of rail and that the larger islands would have had several species. Very few remain—hence his estimate of 2000 extinctions.

Other Pacific bird species have gone, too. All but the most remote islands had pigeons and parrots. Large, meaty, and sometimes flightless, these birds may have been the perfect accompaniment to poi, a purplish paste made from taro root, the staple of Polynesian meals. We must add them to the total. There were also the huge moas of New Zealand, now extinct. And Steadman did not include the smallest bird species because their bones are rarely preserved.

It took just a few thousand years for the Polynesians to colonize the Pacific; their superb navigational skills brought them to even the most remote islands. Armed only with tools of wood, stone, and shell, they exterminated several hundred species. Steadman would argue a few thousand. Let's take 1000 as a compromise figure.[8] It comes to 10 percent of the world's bird species. The extraordinary conclusion is that double-digit extinction percentages are part of Earth's recent history, not merely wild speculations about its future. These numbers mean that the rate of extinction was an unnatural one species every *few years*—not the expected one species per *century*.

The story is not over. James Cook found the Hawaiian Islands in 1778. Trade and colonization followed within a generation. The newcomers cleared forests and introduced cattle and goats, which, like the pigs, destroyed native plants, species as unprepared for large mammalian herbivores as the birds were for rats. Today our only records of 18 species of birds are the specimens collected by nineteenth-century naturalists, and they missed some species. In these islands alone—just one island group—there has been one extinction about every decade for the last 200 years.

What remains of the Hawaiian birds today? Have we put the lessons of the past behind us? With our better understanding, are extinctions a thing of the past?

When Is a Species Extinct?

A small group of ecologists stands in the forest on Kaua'i. It was a long, wet hike to get there, but their reward is a marvelous song, flutelike and varied. They catch glimpses of the bird as it moves through the forest—a male 'o'o'a'a, black with yellow tufts. By accident, one of the group plays the bird's song from a tape recorder. The bird responds immediately and excitedly to his perceived rival, perhaps because he hasn't done so for so long. The group feels ashamed for tricking the bird like this. He is the last one. Months later, on another visit, a playback does not work: The forest is silent. "EX" for extinct will be entered beside the name of 'o'o'a'a in the global catalog.

Across the Hawaiian Islands are a dozen species so rare that there is little hope of saving them, even if they survive. Try finding a po'o uli, or a nukupu'u, a Bishop's o'o, or a Maui 'akepa. That is what my students and I were trying to do in our explorations described at the beginning of this book. Alas, we failed during all those cold days when our clothes never completely dried out. I recruited two postdoctoral students, Paul and Helen Baker, to look full-time. They had excellent credentials, and I knew they would be hardy. Like other northern Britons, they grew up in a climate of cold, continual mist, and rain not unlike upland Maui.

After two years of effort, they found three po'o uli. Two are known to be females. At the time of writing, the sex of the other one is unknown. What is the chance of this species going extinct? Heads, the third bird is a female and the species is gone. Tails, it's a male, but we have only the remotest chance of bringing a pair into captivity to breed. Small birds like these rarely do well in cages. Get ready to add another "EX" to the catalog.

That catalog is called *Birds to Watch 2;* the "2" means that it is an update of an earlier version.[9] It is compiled by Nigel Collar, Michael Crosby, and Allison Stattersfield, all from Bird Life International outside Cambridge, England. A bird gets an "EX" only when it is gone for certain. They will wait a couple of decades before closing the files on the species I have mentioned. Birders are an optimistic lot.

The catalog's message is clear: Species are lost at the rate of about one per year. That is about 100 times faster than the benchmark odds of 1 in 1 million—one extinction per 100 years among the 10,000 species of birds. These are species we have already killed. What about the ones we have fatally wounded?

The Fatally Wounded

The catalog's list has other letters beside the species' names. "CR" stands for critical, "EN" for endangered, and "VU" for vulnerable. Combined, these are the "threatened" species and there are more than 1100 of them. What will be their fate? How long will be the destined hour of species in the new century?

Nigel Collar and his colleagues define "threatened species" as those likely to become extinct in the "short to medium term." What exactly does this mean? Decades? A century? Before I answer, what does this phrase mean for average species' lifetimes? Suppose that it would take as long as a century before all these 1100 species became extinct. On average, 11 species per year would be lost—a rate 1000 times greater than the benchmark of one species per century. Is it credible to suppose that 1100 species of birds would become extinct within this new century? Could the extinction rate be that bad?

Some species, by good luck or our good management of them, will survive the century, and so perhaps fewer than 1100 will become extinct. Yet these are the species that human beings have harmed already, not those we will harm in the future. As we save some species, yet more will be fatally wounded.

Of these 1100 species, nearly 200 are in the "critical" class. These include the Hawaiian species mentioned earlier and the two species on the catalog's cover, the nukupu'u—seen only sporadically over the last few decades—and the akialoa—a species that has not been seen in 50 years of intensive surveying. Many of the 200 species are almost certainly already extinct.

The critical class is defined by populations numbering fewer than 50 individuals or by fewer than 250 individuals combined with a steep decline in numbers. How long could such species last? Two long-term studies from Britain show why the answer is "not long." The first one is a remarkable 1000-year study of the fate of a single, rare population. It shows how such populations risk unavoidable vagaries of sex—the diffi- culties of finding a suitable mate or most of the young of a generation being of the same sex—and of death—most of the individuals dying before reproducing. British ecologists call this long-running experiment "Royalty," who, by their own choices, constitute a small population and provide a uniquely detailed record of disasters, tragedies, and the importance of occasional good luck.[10]

Picking up the story in the sixteenth century, the ecologists' version of British history goes like this. Henry VIII went through half a dozen wives to produce one son and two daughters, none of whom reproduced. The Tudor dynasty ended. In the seventeenth century, Charles II was the prolific father of his country, but not of any legitimate heirs. His brother took over, mar- ried a Protestant commoner, divorced her, then married a royal of an unac- ceptable religion and was promptly ousted. The commoner's daughter, Queen Anne, started the next century by losing every one of her children soon after birth.

George III may have lost half the North American colonies (Americans always forget Canada, Bermuda, and the Caribbean), but he had four sons, born within 9 years. Two became kings. William IV married but had no heirs. The other, George IV, almost single-handedly terminated royalty in the early nineteenth century with his bigamy and otherwise bad behavior. His one legal daughter died during childbirth along with her baby. The fourth son left Britain to rescue a foundering population of minor German royals, while the third had a daughter, Victoria. Victoria became queen, and proved to be prolific and respectable, well able to make up for Uncle William's demographic and social errors. Her progeny repopulated royals across the continent. In the twentieth century Edward VIII was unable to find an acceptable mate.

Only rarely in the last five centuries has this small population consis- tently passed the title from parent to child for more than a couple of gener- ations. Consistent "heirs and spares" are scarce. The population has to be rescued by immigration from other populations of royals. It is exactly the same for other rare species, except that for many there is no possibility of rescue; there are no other populations.

The second study involves the fate of populations of birds on a dozen small islands off the coast of Britain.[11] These are favorite places for avid birdwatchers, and some islands have been visited annually for half a century or more. For dozens of different species, the records show the year when the population first bred on an island, how many pairs nested in subsequent years, and when the population foundered. Each year is a game of chance: Will the progeny survive until next year, will those that survive be of the same sex, and, if so, will an immigrant of the right sex arrive? If so, will the potential mate be a compatible one? Small populations can stay in this game for a while, given luck, but not forever. Most of the populations die out within a few decades. So we know that while some of the 200 critically endangered species will have a run of luck, it cannot last. Others will have bad luck and be gone within a few years.

But what about the other 900 species in the catalog that are not quite so rare? Populations of a few hundred to a few thousand individuals do not suffer the demographic problems of royalty and the species on small islands. They are, however, subjected to a broad range of environmental vagaries— cold winters, droughts, disease, food shortages, and so on—that cause all populations (including human ones) to suffer occasional sharp drops in numbers. Within a century or so such natural disasters can drive even the largest populations of threatened species to levels so low that recovery is unlikely or impossible.[12]

There are even relatively common species whose fate is already sealed. Earlier chapters have shown how quickly we are destroying the forests and drylands that are home to many species. It is a topic to which I return in Chapter 13.

There is another major reason why common species can go extinct. Humans have carelessly moved species around the world, either by accident or by design. Such introduced species eat or compete with native species; they also carry diseases against which native species have no immunity. All this can happen quickly, and, undoubtedly, many species will be lost to introduced species that cannot be anticipated. For example, no one considered the birds on the island of Guam to be in danger 30 years ago. An introduced snake, the brown tree snake, has eliminated all the island's birds since then. Were the snake to reach Hawai'i, all of its birds would be at risk. Single snakes have already turned up at airports in Hawai'i; they creep aboard airline flights from Guam.[13]

Summarizing these facts is simple. Birds are already going extinct 100 times faster than they should. Our past actions have already fatally wounded so many other species that the fatalities will soon occur 1000 times faster than they should. Our current activities will harm even more species.

I have given too many seminars not to know what you are thinking. The dodo was an island bird. It is a universal symbol of stupidity and its fatal consequences. According to *The New Shorter Oxford English Dictionary*, "it became extinct." Not "we bludgeoned these defenseless birds into oblivion," please note. Island birds had it coming to them, the *OED* implies. Perhaps birds are all dodos. Perhaps all island species are, or just birds on islands. They lack the "right stuff" and deserve to go extinct.

Meet the Wainwrights

A Gary Larson *Far Side* cartoon expresses the idea best. A father casually justifies to his son the events next door, where wolves are dismembering the neighbors, the Wainwrights. That's OK, the father assures him: You shouldn't mourn their passing. They were nice people but "weak and stupid." The Wainwright hypothesis is that extinct species were weak and stupid, and we should not feel sad about their passing either.

Island species are unusually vulnerable to extinctions. Not only have the Pacific Island birds suffered but so have the plants and invertebrates. Out of a total of nearly 1000 plant species, the Hawaiian Islands have lost more than 80 since botanists first arrived to describe them. Many more are threatened, and 133 species have fewer than 100 individuals.

Ten of these species are down to single individuals in the wild. I have looked at plants knowing that I could uproot them in a minute and drive a species to extinction. Not that I would, but a hungry pig, goat, or cow could hardly be expected to care. How many species disappeared as victims of the Polynesians, the rats and pigs they brought, or the fires they set, we may never know.

The specimen cabinets of the Bishop Museum in Honolulu contain hundreds of local varieties of land snails eaten to extinction by a predatory snail. The predator was deliberately brought to Hawai'i in a futile attempt to control another introduced snail. In spite of warnings from biologists, the experiment was repeated across the Pacific with the same devastating results. In much of French Polynesia, every native snail disappeared from the wild.

Species are being lost on other islands as well. In the last 300 years, Mauritius, Rodrigues, and Réunion in the Indian Ocean lost 33 species of birds, including the dodo, 30 species of land snails, and 11 species of reptiles. St. Helena and Madeira in the Atlantic Ocean have lost 36 species of land snails.[14]

You might deduce from this that all island species may be Wainwrights. Not so. Mainland species suffered as well. Consider the fynbos, an area at the southern tip of Africa rich in plant species. Of a total of 8500 species, the fynbos has lost 36 in the last century. Over 600 species are threatened with extinction from the loss of habitat and introduced species.[15]

Or consider Australia. Most Australians live in coastal cities, many having never visited the sparsely populated dry interior they call "miles and miles of bloody Australia." Yet of the global total of 60 extinctions of mammals in the last 200 years, 18 species were in Australia, about half of them from the dry interior. Those are from a total of nearly 300 native Australian species.

You do not need to travel to remote locations to meet Wainwrights. Of the nearly 300 species of North American freshwater mussels and clams, 21 are believed to have gone extinct since the end of the nineteenth century.[16] Another 120 species are threatened. Forty kinds of freshwater fish in the United States, Canada, and Mexico have become extinct in the past 100 years from a total of nearly 1000 species.[17] Wainwrights are everywhere.

Worldwide, only a few groups of plants and animals have catalogs as complete as the one for birds. So it is not easy to determine which kinds of species are most at risk. For North America, however, The Nature Conservancy has made a special effort to inventory as many groups of species as are reasonably known.[18] There are 13 groups. The study counts extinct and probably extinct species and those in immediate danger of extinction. Adding up all the species in these categories gives the following percentages, listed in increasing order:

Butterflies	4 percent
Birds	6 percent
Mammals	6 percent
Tiger beetles	6 percent
Reptiles	7 percent
Dragonflies and damselflies	8 percent
Ferns	10 percent
Conifers	13 percent
Flowering plants	16 percent

Freshwater fish	21 percent
Amphibians	23 percent
Crayfish	37 percent
Freshwater mussels	43 percent

Birds are proportionately the second least endangered group. Far from being the Wainwrights of the world, they are of the best stuff!

Are all these species the weak and stupid ones that the Larson cartoon suggests? Not at all. High rates of extinction are typical of continents and islands, of forests and drylands and rivers, of plants, and invertebrate and vertebrate animals. To the extent that we can squint at these numbers and deduce a pattern, it is to determine that freshwater species are at greatest risk. Given the impact that humanity has on rivers, this should not come as a surprise.

The Wainwright cartoon pokes fun at our preposterous justification that extinct and threatened species had only themselves to blame. Larson also has a cartoon about dodos. It shows them as uniquely advanced birds, a cultured, literate, civilized species. Humans arrive and kill them all. Larson gets far closer to the truth than does the OED.

An Unnatural Minute, Not a Destined Hour

In the metaphor of the poem *Rubaiyat of Omar Khayam* that gives this chapter its title, species should appear, live their destined hour, then pass away.[19] The senator's question is whether the law tries to prevent extinctions that are perhaps inevitable and thus should not warrant attention.

The fossil record suggests that most species live 1 million years or more. The DNA evidence provides confirmation. So the destined hour for species will be measured in the millions of years. At this rate only 1 species in 1 million or more should encounter its natural end each year.

The unnatural number is already closer to 1 species in 10,000—a number 100 times larger. Looking at the numbers of species that we have already wounded and making the grim assessment of how long they are likely to live, we find that the number is closer to 1 in 1000, which is 1000 times larger than the natural rate. Humanity's impact has reduced species' lifetimes from a metaphorical hour to a minute, and it soon may be a matter of seconds.

These reductions in species' lifetimes are not only large but ubiquitous. They happen throughout the planet's diverse ecosystems. Although the extinction of large, charismatic birds and mammals catches the public's attention, species as diverse as flowers, land snails, freshwater bivalves, and fish are also going quickly.

Is there anything special about these species? Only that they were collectors' items a century ago, so we are well informed of their fate. Extinction rates in these diverse but familiar groups are typical of the many millions of species that taxonomists have yet to enter into their catalogs.

If every group of species *and* every place are affected, does this mean that every group of species *in* every place is affected? We might expect extinctions when humanity collides with species. Although human impacts certainly explain all the case studies mentioned, peculiar anomalies do occur. There are so many extinctions on the Pacific Islands and in Australia—hardly the most densely populated part of our planet—and yet so few in densely populated Europe. Why should this be so? The answer is that some species in some places are more vulnerable than similar species in other places. It is a matter of where the species are—the constraint of space—and that's what I'll discuss next.

13

Nature's Eggs in Few Baskets

Charles Darwin left England in early 1831 and arrived in Brazil a month later. He was soon writing to his father, peppering his letters with descriptions such as "striking," "luxuriant," "no person could imagine anything so beautiful," and "exquisite glorious pleasure." The diversity of life "bewilders the mind. If the eye attempts to follow the flight of a gaudy butterfly, it is arrested by some strange tree or fruit; if watching an insect, one forgets it in the stranger flower it is crawling over. . . ." In the forest, he enthused about "the elegance of the grasses, the novelty of the parasitical plants, the beauty of the flowers, the glossy green of the foliage . . . the noise of the insects." He went on and on like this for pages.[1] When escaping an English winter, the tropics are always a sensory shock.

These days, visitors from the northern United States, Darwin's fellow Brits, and other Europeans—"snowbirds" we call them—arrive in Florida in January and snorkel off its coral reefs. They clamber back into the boat and exclaim in a dozen different languages about the fish and corals they have seen. More adventurous cousins visit the Amazon, or Costa Rica, or even wilder places. Relatives back home have only their carefully tended houseplants and tanks of tropical freshwater fish to bring color to their lives.

Where the Species Are

The first question to answer in this chapter is, Are species numerous everywhere or only in special places? Darwin's gushing prose proclaims, "the tropics!"—their land, their waters, their oceans. Does this mean that the planet will lose more species in the tropics as a simple consequence of its having more species to lose? To answer this second question, we must ask

217

whether all species are equally vulnerable to human impact. They are not. The most vulnerable have unusually small geographic ranges. Are vulnerable species—nature's eggs—simply scattered at random or are they concentrated in special places?

In this chapter I show that vulnerable species are concentrated in special places we call *hot spots*. Nature has put her eggs in a few baskets. Most, though not all, hot spots are in the tropics. South Africa's plants, North America's fish and freshwater clams, and Australia's mammals are exceptions. Everywhere in the tropics is not a hot spot. Finally, I show that where the human impacts described in Parts 1 and 2 of this book collide with the hot spots, species become extinct in unusually high numbers.

About two-thirds of animal and plant species live in the tropics. This frequently quoted high percentage is a rough guess, one that comes from pulling open the drawers of museum cabinets and checking off species. We already learned that most species are known from only one specimen and that perhaps 10 times this number are not known at all. The species we know best are from the world's species-poor temperate regions. "Two-thirds" may be too low a guess, for the species we have yet to describe will be mostly tropical.

Even the tropical species in museum drawers are from places we know best. Remember Roger Kitching's lamenting how little he knew about insects and mites beyond a short walk from "O'Reilly's," the lodge in Lamington National Park's rainforest? The same problem applies to many other rainforest sites and lots of other tropical environments. One explanation is common to all. Ecologists like to stay close to a hot shower and a cold beer. And each site offers a particular favorite cold beer: Castlemaine at O'Reilly's; Imperial at La Selva in Costa Rica; HB at the Smithsonian's station at Barro Colorado Island, Panama; Antarctica at Reserva Ducke, the site near Manaus, and on down the list. All but the Brazil site have hot showers.

Who knows what happens away from these sites? If we have no idea where the wild things are, then our ability to predict human impacts is going to be limited. Do we know anything for certain about where species are?

Well, we know where birds are. Birders (who are overwhelmingly male) like their beer even more than the average man. Yet they are even more passionate about their life lists. Not only are there associations for expressing fondness for birds, but ones that publish just how long birders' lists are. To make the cut, you must venture forth, beyond where the long hikes shake the cans of now-warm beer to the point of bursting, and into the remotest corners of the planet. Steamy rainforests, cold high mountains, the driest deserts, all must be explored without any heed of creature comforts and regardless of dangers. Bird watching is an extreme sport.

For birds, and only for birds, we have a near-complete catalog of the planet's 10,000 species. It is written by Charles Sibley and Burt Monroe and called *Distribution and Taxonomy of Birds of the World*.[2] We also have *Birds to Watch 2*—the book described in Chapter 12 that lists which species are extinct or likely to go extinct soon. Across almost every part of the planet we have reasonably complete mapping of every one of their geographic ranges.

Birders' explorations readily confirm that more species are in the moist tropical forests than anywhere else. Consider some basic bird-watching facts. Europe has fewer than 500 species of breeding birds; North America, from Alaska and Canada south to the Mexican border, has about 800. Travel south and these continental benchmarks are easily exceeded by individual tropical countries. In Africa, the countries of Kenya, Uganda, and Tanzania each have about 1000 breeding species; in South America so, too, do Ecuador, Colombia, Peru, and Venezuela. Brazil has even more.

These countries are large, so their totals paint the picture of biodiversity in broad strokes. To answer this chapter's questions we need finer detail. For birds we have them in the form of maps of the ranges of where each and every species nests. To provide an example, I chose the New World: North and South America and its perching birds or passerines. The *passerines* are an order of birds and are generally the smaller species, such as flycatchers, swallows, wrens, thrushes, warblers, and tanagers. Crows are the largest passerines. Larger species, such as herons, ducks, hawks, parrots, and pigeons, belong to different orders. The Americas are home to more than 2000 species of passerine, nearly a quarter of all the world's birds.

Lisa Manne, who joined my research team to complete her dissertation, was a refugee from mathematics. She had decided that life was more interesting than equations after volunteering one spring for fieldwork in the Florida Everglades. Her special talents were in computer programming. To her fell the task of writing the computer code that would organize and analyze the millions of individual entries detailing which bird species were found where across the Americas[3] (see Map 12).

Lisa's first question for her doctoral dissertation was the simplest one: Which areas have the greatest number of species? Her computer program counted the number of bird species that are found nesting in each block of 1 degree latitude by 1 degree longitude across all of the Americas.

The map is shaded by divisions that double; were it not so, North America would appear as one uniform color with few species found there. Starting from the north, the tundra of northern Alaska and Canada has fewer than 32 species per block. The numbers double as we reach boreal forests; these can house as many as 64 species. So, too, can the central prairies, but the deciduous forests to the east and the coniferous forests to the west have between 64 and 128 species. The penultimate doubling to 128 to 256 species arrives on cue at the northern tropic in Mexico. The darkest shading shows areas of more than 256 species; they are in the Amazon basin within a few degrees of either side of the Equator and along the coastal strip of southern Brazil.

Traveling southward, the pattern is reversed. The species numbers decline, dropping below 128 species south of the southern tropic, almost on cue again. More species are on the Atlantic side, which is forested, than on the Pacific side, the Atacama Desert. By the southern tip of Chile and Argentina the species numbers have dipped below 32. Islands such as the ones in the Caribbean and the Galápagos have few species compared to the adjacent mainland.

Two simple patterns emerge: More species are in the middle than at the two ends; and islands have fewer species than comparably sized areas on the nearby mainland. Why islands have fewer species is an easy question to answer. It is difficult for some species to reach islands because they have to fly over water. Birds and bats can fly, but not all can fly well. For almost all other species, getting to an island is an unlikely event.

Why are more species in the tropics? Ecologists have long debated this question, proffering dozens of hypotheses, each couched in the arcane jargon of the profession. All may be true, but I have my own favorites. One is simple, the other sublime.

The Importance of the Middle

Robert Colwell, an ecologist from the University of Connecticut, and I were sitting in a bar drinking Antarctica and excitedly drawing lines on paper napkins, the preferred medium of serious scientific discourse worldwide.[4]

The simple explanation for more species in the middle, the tropics, is that they cannot be anywhere else! Species with large geographic ranges have to be nearly everywhere or else they wouldn't have large ranges. And this fact alone means that they must be in the middle. Grab a paper napkin or two and draw a line from northern Canada (N) to the southern tip of South America (S):

```
                        Equator
North _____|_____ South
```

The first species, let's call it A, has the largest range. It could be as northerly as possible:

```
                        Equator
North _____|_____ South
AAAAAAAAAAAAAAAAAAAAAAAAA
```

or as southerly as possible:

```
                        Equator
North _____|_____ South
        AAAAAAAAAAAAAAAAAAAAAAAAA
```

or anywhere in between. It does not matter where you choose to put A; it has to be found across the middle.

So pick a range at random, somewhere from as southerly to as northerly as possible. Then add another species, B, with a slightly smaller range. Place it at random, too.

```
                        Equator
North _____|_____ South
      AAAAAAAAAAAAAAAAAAAAAAAAA
        BBBBBBBBBBBBBBBBBBBBB
```

Now pick another with a slightly smaller range, and another, until you get thoroughly bored. You will find that only species with smaller ranges can be choosy, but some of them will be in the middle, too.

Here is what I came up with:

```
                        Equator
North _____|_____ South
      AAAAAAAAAAAAAAAAAAAAAAAAA

        BBBBBBBBBBBBBBBBBBBBB

   CCCCCCCCCCC

            DDDDDDDDDDD
```

I have a low tolerance for boredom, but I can already see the pattern after only four tries. Computers, however, handle boredom really well. A computer can do what I have done over and over again, hundreds of times, all without complaint. So, Rob unpacked his computer; it was the only thing that was not complaining in the bar. (Its clientele is stunned, incredulous. Not only has France just beaten Brazil in the World Cup, but it was not even close. With no wild celebrating in the streets, playing computer games was the only entertainment Río de Janeiro had to offer us.)

Rob's computer picked hundreds of hypothetical species ranges just like the few we'd drawn on our napkins and rapidly produced maps of areas that should have the most species. There were also more species in the tropics because they are in the middle of the two continents.

If this were all there was to the story about where species are, then we were in the wrong place. Río is just inside the tropics, the southern edge of the middle. Manaus, in the middle of the Amazon, has many more species. Why we were in Río and not Manaus had everything to do with the species that have the smallest ranges. Where can you put the many species that have small ranges? Species with small ranges do not have to be in the middle. Try species X. It could go anywhere!

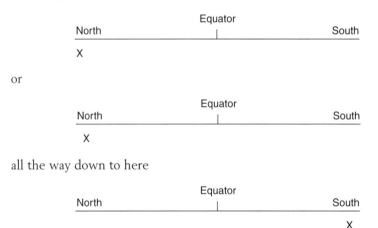

or

all the way down to here

So if all species had tiny ranges, the tropics might not have more species than the higher latitudes.

Are the species with small ranges just anywhere? Nobody answers that faster than the birder who wishes to demonstrate his top status by having a longer life list than his rivals. Birders want to know which species are the *endemics*, or species found within a limited area and nowhere else. Bird-finding guides highlight these species by underlining them or putting an "E" in the margin whenever they are mentioned. "Don't go home without seeing these," the text exhorts, "for you won't see them anywhere else."

Small Ranges, Extinctions, and Endemics

You should know three facts about endemics. First, if their habitat is destroyed, it is not just birders who won't see them. Neither will anyone else. Second, there are more endemics than you have any reason to expect. Finally, they are not strewn randomly about the planet. They are concentrated, and the concentrations—the hot spots—often lie directly in the path of humanity's advance.

The first fact is the obvious one. The smaller a species' range, the more easily humanity can harm it. It is easier to clear cut, overgraze, overfish, degrade, or otherwise despoil a smaller area than a larger one. By a huge margin the species that have become extinct in the last century had small ranges. By a similar huge margin, those currently threatened with extinction have small ranges.

Before leaving for Río, I had asked Lisa Manne to do the requisite calculations. She retreated to her office to write more computer code and returned within the hour with her answer. "Most of the threatened birds of the New World have ranges smaller than 60,000 km^2," she told me.

Small is vulnerable. On the birders' lists "E" is for *endemic;* it might as well stand for *eggs*—Nature's most vulnerable species. Find a species with a small range and odds are that it will be threatened with extinction.

Is small range size all that matters? Almost, but not quite. Some species with large ranges are threatened. Whales had the world's oceans in which to hide. Like Captain Ahab, we relentlessly hunted them down to the last individuals of some species. The species with large ranges that are still threatened with extinction are nearly all species that humans hunt for their meat, their skins, or, as with parrots, their company.

The second fact is not obvious. Why are there more endemics than you would expect? Everything in nature varies, and we all share an intuitive understanding of that variation. "On average" means that not everything of whatever-it-is-to-be-measured is of exactly the same size. There is variation. Yet we talk about an average with the confidence that everything will be similar to that average, if not exactly identical to it. The less similar something is to the average, the less likely we are to encounter it. Most people are of near average height, not much taller or much shorter.

Sometimes that confidence fails us: Average values do not always do a good job of describing some things. Range sizes are a case in point. The passerine birds of the New World have an average range size of about 2 million km^2. Yet most of the ranges are not anywhere near this large: Half the birds have range sizes of less than 600,000 km^2. Many species have small

ranges and only a few species have really big ones. These data skew the average. Few birds have average range sizes. Nature has produced a lot more eggs
than you might expect.

So where has she put them? Endemics could be anywhere, but they are
not. I asked Lisa to come to my office again. "Please map where the species
with small ranges are. Let's pick endemics whose ranges are less than
100,000 km^2." She retreated to her computer and put on her headset, pre-
tending to listen to music but really wanting to blot out any more of my
requests.

A 1-degree-longitude by 1-degree-latitude block is approximately
10,000 km^2. The species I'd requested would fit into ten or fewer such
blocks. Any set of ten blocks could hold one or more of these endemics. The
southern tip of South America could and so, too, could the islands of
Ellesmere, Baffin, and Victoria off the coast of northern Canada.

Again within the hour Lisa returned with a map (see Map 13). Almost
to prove the point that species with small ranges can be anywhere, one
species is stuck at the southernmost tip of the Americas. There are no
species in Canada. One species is close: stuck in Michigan, closer to the
North Pole than any sensible species should live. There are a couple in Texas,
and a couple just south of the southern tropics in Chile and Argentina. But
that is it for the areas outside the tropics; all the rest are tropical.

Even within the tropics, species with small ranges are not everywhere.
The entire Amazon basin and the drier forests to the south and east hold
very few of them. Yet this region has the highest number of species on Lisa's
first map of all species.

A major concentration of endemics is along the continent's western rim
from the Sierra Madre of western Mexico, through Costa Rica, Panama,
Colombia, Ecuador, Peru, and Bolivia. Another concentration is along the
southern Atlantic Coast of Brazil, with its peak in the state of Río de Janeiro.
There are the offshore islands—the islands in the Caribbean and, off the
west coast of Ecuador, the Galàpagos. Islands have few species but are rich
in endemics.

Rob and I weren't sitting in a bar in Río just waiting to join a million
inhabitants in a night-long samba party if the Brazil team had won. We were
between trips to visit the few remaining fragments of Brazil's coastal forests
and their numerous unique species—their endemics. Within a day's drive of
Río were 200 species of bird found *only there*. We called it a night and walked
back along the empty streets to our hotel near the beach at Botofago Bay.

Charles Darwin had stayed in a cottage on the bay in 1832. Elegant
houses were scattered along it. Steep forested hillsides rose inland—so steep

that they still hold forest. He loved the cottage. But his letters, brimming with enthusiasm only weeks earlier, had changed. Before leaving Bahía a prick on the knee had become inflamed, and he was confined to a hammock for 8 days. Some of the ship's crew had been "disagreeable" and were being sent home. On landing, the cutter was swamped in heavy seas, Darwin was knocked on the head, and his books and equipment plunged into the water. From Botofago, he ventured inland on a miserable 2-week journey. He was sometimes ill, often hungry, could not undress for days, and was terrified by not being able to speak the language to get help. He was appalled by the slavery he saw. It rained torrents. Other crew members were dying of fevers, quickly and violently.

In letters home from remote places we all try to conceal the bad news from our families. Darwin wrote to his sister Caroline and did not do a good job. He knew the infected knee, the boating accident, the trip inland, and malaria or other fevers could have killed him quickly, far from home. He was probably as terrified as I have been from time to time. He completely missed the endemics. But then he had no reason to expect species with such small geographic ranges. They are as scarce in Europe as they are in North America.

Three and a half years later, the *HMS Beagle's* mission of mapping South America was complete. From the continent's Pacific Coast the preferred route home is westward, to avoid Cape Horn. The *Beagle* stopped for just a few weeks in the Galàpagos. Its dense concentration of endemics off the continent's west coast is evident on Lisa's map. You know the rest of the story. He did not miss the endemics here.

For years, his diaries were filled with careful and detailed descriptions of the species he found on his voyage. Yet in the Galàpagos he still mixes specimens from different islands. Since the islands are close together he has no concept that species could differ across such short distances. Island residents tell him again and again that he should not mix his specimens. Finally, he recognizes his mistake and marvels at what he discovered. Each island has a unique species of mockingbird, similar to each other, but separate, even though the islands are close. The same applies to giant tortoises and small black finches. They are "aboriginal productions," he wrote—*this is where species are born*, he realizes.

He made it home the next year and compiled notebooks to express his thoughts. Then he waited, and then he waited some more. Twenty years later Alfred Russel Wallace got the same idea and from exactly the same experiences. Looking at endemics on the islands of Southeast Asia he, too, realized that areas rich in endemics are species nurseries.[5] Wallace wrote to

Darwin with his idea and Darwin could procrastinate no longer. They wrote short papers back to back in a scientific journal. Not much happened. The following year Darwin wrote *The Origin of Species*. It sold out in a day and radically changed our thinking.

Why on the islands of the Galàpagos? Why along the Andes? Why the coastal forests of Brazil? Why in northern Michigan, for pity's sake? Why are species with small ranges where they are? They could be anywhere, but they are not. They are not even where there are the most species.

These are not the questions Darwin answered. Despite its title, *Origin* is about how species change, not how they are born. But the idea of their birth gave Darwin and Wallace their crucial insights. A century and a half later we still have an incomplete understanding of why species nurseries are where they are. Nor are these the questions I shall answer. For I'm concerned about species' deaths.

Lisa's third map shows where the threatened species are, where species are now dying. It looks like her second map (Map 13), so I have not included it in this book. Río de Janeiro has the highest concentration of bird species on the brink of extinction in the New World. "E" is for *endemics*, and "E" is for species that are about to become *extinct*.

These American patterns are not unique: They repeat worldwide. Some places have almost no endemics: Continental Europe has only two. Other places are teaming with endemics. The Hawaiian Islands once had nearly 150 endemic birds. The Southeast Asian islands of Sulawesi, the Solomons, New Britain and neighboring New Ireland, the mountains of New Guinea, Mindanao in the Philippines—each hold more than 50 endemic species.

The World's Hot Spots

These areas, plus several from the New World mapped by Lisa Manne, are the hottest bird hot spots. None of these areas is large. Central New Guinea and Sulawesi are slightly larger than England and Wales combined, and Mindanao is about the same size.

It is not just birds that have hot spots, of course. The peripatetic Norman Myers and his colleagues at Conservation International in Washington, D.C. reviewed the global patterns for plants, amphibians, reptiles, and mammals.[6] We do not have the detailed maps for these species that we have for birds, but the global patterns still show the same general patterns.

For the 300,000 plant species, most are in the tropics, islands have fewer species than continents, and some special places are very rich in endemics. For example, Central and South America have about 85,000 species compared to North America's 17,000. The isolated Hawaiian Islands have fewer than 1000 plant species. Similarly, Africa has about 45,000 species of plants to Europe's 12,000. As for birds, the patterns for the endemics don't always match the patterns of overall species numbers. The Hawaiian Islands may not have many species, but over 90 percent of them are endemic.

Some of the plant species patterns match those for birds. Eight thousand plants are found only in the Atlantic Coast forests of South America; another 20,000 are found only in the tropical Andes. But not all the patterns match. Outside the tropics, the Mediterranean basin has 13,000 endemic plants; so botanists do go to France, even though that area is a letdown for bird-watchers. Botanists also find similar kinds of vegetation elsewhere: California has 2000 endemics and southwestern Australia has 4000. Botanists, as you may have surmised, can be a more sophisticated crowd than beer-guzzling birders, one more likely to appreciate a good Rhône, zinfandel, or chardonnay. And a good pinotage, for that matter. Southern Africa is spectacularly rich in endemic plant species. The Cape, a mere 18,000 km^2 of the southern and southwestern tip of the continent, holds more than 8000 plant species of which a staggering 70 percent are endemic.

All told, Myers and his colleagues find that nearly half of all plant species dwell within 25 hot spots, a combined area of only 17 million km^2 of the land surface. Within what remains of the natural habitats of these areas are also 30 percent of the world's birds, 30 percent of the mammals, 40 percent of the reptiles, and more than 50 percent of the amphibians.

It isn't only the land that has hot spots. North American rivers are the world centers for freshwater mussels. In one family of mussels, half of all known species are in the eastern United States. With few exceptions the freshwater mussels of eastern North America exist nowhere else in the world. Chapter 7 documented how extensively those rivers have been modified, and over 40 percent of North American mussels are extinct or nearly so as a consequence. The Amazon forest, shrinking year after year, nurtures a maze of individual rivers that hold a quarter of the world's species of freshwater fish.[7] The dazzling variety of marine fish and invertebrates on coral reefs (including the species of corals themselves) are a source of delight to everyone who dons a mask and a snorkel. Coral reefs are almost entirely tropical; in a few places they stretch to 30°N and 30°S. They are home to the majority of the world's marine fish. Callum Roberts and his wife, Julie Hawkins, at the University of York, in England, have scrutinized which fish (and corals and snails, for that matter) occur where. The same pattern of hot spots holds for these marine groups. A large fraction of the species is concentrated into a set of very small areas. The waters around the many islands in Southeast Asia house a large proportion of the world's reef fish.[8] On land, in freshwater, and in the ocean, Nature has put her most vulnerable species into a small number of species places. To pursue this metaphor further, we would ask: Have we protected these special places? The answer is worse than "no." We actually singled out these places and harmed them disproportionately. Averaged over the entire land surface of the planet, Chapter 6 tells us that we have destroyed about two-fifths of the plant production. Yet averaged over the hot spots, our activities have destroyed closer to four-fifths of the natural ecosystems.

Human impacts do not single out natural habitats at random. We have high-graded the planet, taking the best bits. In South America, for instance, human population is concentrated along the Atlantic Coast, exactly where concentrations of the species with small ranges are. In the oceans, we have made a major impact on coral reefs, areas with the most biodiversity. And so on. It is as if we deliberately chose to hit the hot spots the hardest, as if we malevolently chose to destroy biodiversity as quickly as possible.

Surely humanity does not completely destroy habitats. Don't we leave something behind? "Not much" has to be the answer for many parts of the world—Cebu in the Philippines, for instance. Nonetheless, there are some parks and other protected areas. How many species will survive in them?

How Many Survivors Will There Be?

Worldwide, about 6 percent of the land surface is protected. This average is misleading in two ways. First, many of the parks are what we call *paper parks*, like the one on Cebu in the Philippines. Here protection on paper does little to protect the park's species. Second, this 6 percent is mostly in the cold tundra atop mountains or in the Arctic, beautiful areas but not rich in endemics. Approximately 2 percent of tropical areas that are rich in species are protected.

Suppose that by enormous effort we could raise the figure to 5 percent and, moreover, we could strictly enforce species protection. How much of the planet's biodiversity would we save? If we save 5 percent of the land, will we save only 5 percent of its species? With luck, the answer is closer to 50 percent, a number that needs some explanation.

The land we save will often be "islands" of forest or other remnant habitats surrounded by a "sea" of farmland or, more likely, poor grazing land. How many species a real island—land surrounded by sea—contains depends on its size. Larger islands have more species, and we know exactly how many more are birds and butterflies, and flowers and fungi, and indeed an unusually broad range of organisms.

To get such answers ecologists must explore islands surrounded by ocean. They must visit islands of different sizes and count the numbers of birds, butterflies, or whatever takes their fancy. Better results come from islands that have lots of species, which means tropical islands, in practice. Yes, I know it is tough having to go to the Caribbean or Polynesia or the Seychelles, especially in January or February, when one's colleagues in northern climes have to stay behind to run their laboratory experiments and then struggle home in ice storms. But we ecologists have the necessary resolve.

We courageous ecologists, returning home with our desperate tales of how an ice-cold Antarctica is not really as tasty as a pint of Newcastle Brown, find a common pattern as reward for our adversities. The data draw out a single curve relating area to species called the *species-area curve*. It is a curve, not a straight line. An island half the size of a larger one will typically hold 85 percent as many species. An island 5 percent the size of a larger one will hold about 50 percent as many species.

That is all well and good, but does the curve work for forest islands surrounded by farmlands? Imagine the experiment one should perform. Take a continuous forest and break it into smaller habitat islands. Allow species to become extinct by keeping the forest fragmented for a while. The extinctions might take several decades or even a century. Check the actual extinctions against the number predicted from the species-area curve. Does this seem like an impossible experiment? Hardly. It is one that humanity has done and is doing repeatedly.

Consider eastern North America, since we already know its forest history.[9] We also know its birds, thanks in part to John James Audubon who was out shooting and painting them 200 years ago. The forest low point was around 1870 when half the area was forested (see Chapter 3). The species-area curve states, "Keep half the forest and keep 85 percent of the species" or, put another way, "Lose 15 percent of the species." Fifteen percent of 160 species is 24.

Audubon shot four species that are now extinct: the Carolina parakeet, Bachman's warbler, the passenger pigeon, and the ivory-billed woodpecker. Another species, the red-cockaded woodpecker, is teetering on the brink. It survives only by energetic conservation efforts. This discrepancy between the 24 and the "4 or 5" bothers critics who write (seriously) that concerns about extinctions are a conspiracy led by Harvard Professor Edward O. Wilson. I have learned more from reading bedtime stories to my children than from these critics. How many bird species would have become extinct if, like the Once-ler of Dr. Seuss's story *The Lorax*, we had cut "the very last . . . tree . . . of them all" from Maine to Florida and westward to the prairies? It's the same as asking how many birds are endemic to eastern North America.

The answer, as my colleague Robert Askins and I have pointed out, is about 30. Eastern North America is a cold spot, not one brimming with endemics. Some 130 of the 160 species are widely spread across the continent, many are found across Canada, and a few are quite common in the deserts of the Southwest and, indeed, south into Mexico. These species would have survived a Once-ler attack. Some of them should have become locally extinct in eastern North America (and some of them did) and then recolonized from Canada and elsewhere as the forest returned. (And some of them did that, too.) Only the species endemic to the area ran the risk of going extinct everywhere.

Applying the species-area curve to the 30 endemics, we predict that 15 percent of them (4.5 species) should become extinct. Exactly between 4 and 5, and exactly the right answer.

Tom Brooks, Andrew Balmford, Nigel Collar (lead author of *Birds to Watch*), and I extended this recipe to birds and mammals across the islands of Southeast Asia and to the Atlantic Coast forests of South America.[10] The deforestation here is more recent, so we counted the fatally wounded species, those thought likely to go extinct in the near future. We confirmed the match between predicted extinctions and species teetering on the brink. In doing so, we confirm predictions of the large fraction of species that will go extinct as we continue to destroy the world's forests and degrade its drylands.

The Future of Biodiversity

Nearly 2 centuries ago, Darwin wrote with delight and wonder about the rich variety of life found in the tropics. Yet on his first stop, the coastal forests of Brazil, he overlooked an extraordinary feature of the species there. It wasn't that coastal Brazil had more species; rather, it had vastly more species within small areas. Research in North America, Europe, or Japan simply does not prepare one for this experience.

Not all hot spots are in the tropics, but most are. Hot spots can be extraordinarily concentrated; thousands of species may be found within an area the size of a U.S. eastern state. Species with small ranges are particularly vulnerable to impacts. Nature has put her eggs in a small number of baskets, and we are in danger of dropping them.

We can even calculate how many. The close match between the numbers of extinctions predicted by the species-area curve and the numbers of already extinct (as with North America) or soon-to-be-extinct species (as in the work of Brooks and Collar) allows simple yet chilling calculations. Myers and his colleagues estimate that only about 2 million of the original 17 million km^2 of hot spots remain. And only about 800,000 km^2 are protected; that is close to 5 percent of the original area.

Five percent of the area will hold only half of the species, which means that half of the 30 to 50 percent of plant and animal species within these hot spots will likely be driven to extinction. The numbers come to between one-sixth and one-quarter of all species, but they underestimate because they do not include the corresponding 50 to 70 percent of species not in hot spots.

Most—perhaps two-thirds—of these remaining species are in tropical moist forests in the Amazon, Congo, and insular Southeast Asia. Ten percent of these forests are being cleared per decade, perhaps more, and certainly the rate is accelerating. Suppose that we manage to protect some of these forests within the next decades. Indeed, suppose that we manage to more than double the area currently protected and save 5 percent. Then we should lose half of these tropical forest species as well. These numbers also work out to be between one-sixth and one-quarter of all species. Adding the two calculations, we could lose between one-third and one-half of all species.

One can play many variations on these number games, but the numbers are probably on the low side. Certainly a few species will survive in the disturbed habitats, and hence make the numbers too high. Countering this tendency is the observation that we drive some widely ranging species to extinction because we hunt them. In addition, alien weeds, insects, and mammals introduced from other continents often thrive in disturbed habitats. They encroach so much on park boundaries that they, not habitat loss, are the main threat to many endangered species.

At the close of Chapter 6 I mused as to whether it was possible to get the answer to the question of how much plant production we consumed on land in a simple way. Recall that it took five chapters to explain how I reached the numbers. This chapter's bottom line is that we might lose between a third and a half of life on Earth as a consequence of our actions. This conclusion follows from three chapters of discussions about what species are, how long they should live, and how long they do live, and where they are found. Is there any easier way of getting the bottom line? Roughly, we lose species in proportion to our human impact. It would be surprising, for example, if we took only 10 percent of the land's production and destroyed 90 percent of its species—or vice versa. There are a lot of details that could lead to such discrepancies. For instance, we could save many species by giving the hot spots special protection, but we do not: Indeed, the hot spots suffer unusual levels of damage. The details that could lead to such discrepancies tend to cancel out over many different places, each with its own unique history of problems. Since I expected our overall impact on land to be about two-fifths, I also expected these numbers on the loss of species.

Epilogue

Like most other ecologists, I travel to beautiful, remote places that are home to peculiar plants and animals and that have few human inhabitants. Most ecologists have an aversion to studying highly modified, densely populated, species-poor areas. It is becoming harder to find the pristine places, though, as the Prologue's helicopter trip documents.

The rainforest tower outside of Manaus afforded an uninterrupted view of the forest's canopy a decade ago. Now the city limits are obvious. Even across the eastern tropical Pacific, when we sailed for weeks without seeing another ship, we saw the floats of long-liners that were over the horizon. Had we caught the sharks we saw—and some other research vessels do—we would have found that many, perhaps most, of them had hooks in their mouths. Even out here one cannot escape human presence and its impacts.

You do not need to be an ecologist to see the extent of human impact. The evidence of human interference in the oceans is as near as any supermarket. I quizzed fishmongers about the availability of cod, swordfish, and herring, learned to cook Chilean sea bass (the latest whitefish to appear on their slabs), and then boycotted it when I learned how quickly the stocks were being depleted. I traveled to Alaska and got to taste wild salmon again, having forgotten what a poor apology for the real thing are the pen-raised fish that now dominate the market.

Earth at the turn of this new millennium is suffering from huge and unmistakable human impacts. Some—the loss of species, certainly, and the loss of tropical forests, almost surely—are about to become irreversible. What does this mean for our future as human beings? What will Earth be like as our numbers double?

233

Will we starve? Of course *we* won't. The United States ensures that poor Mexicans cannot flood across its border, including those people impoverished because the United States drained the Colorado River, a water supply for Mexico. The United States joins with European nations in stationing a huge military force in the Persian Gulf to ensure that gasoline, now cheaper than bottled water, will remain that way. All the evidence points to our nation's intolerance of anything that would even marginally depress the quality of our lives by reducing our access to resources, any resources.

Will others starve? Of course, and millions do so now. Or else they die from a myriad of reasons all connected to living in the wrong place. The number of the have-nots in the world continues to rise steeply. The rich have gotten richer, the poor poorer and more numerous.[1] Will their well-being improve? Others have debated that answer and will continue to do so.

Our well-being is not the issue here. The book has been about what haves and have-nots alike do to the Earth to make a living—what we do to Earth, not what Earth does for us. To the extent possible, I speculated on what we will do in the future. Our impacts will grow as our population increases. Indeed, they will increase at a much faster rate than our numbers simply as a consequence of our having been squeezed into hitherto constrained spaces. The forests yet to be cleared, grasslands yet to be ploughed, rivers yet to be dammed, fish yet to be caught have all escaped our attention. They will be that much harder to exploit sustainably. Even if the human population remained constant, the vast majority of people who cannot afford to read this book have exactly the same aspirations as people who can. Our impacts are a prelude.

Why should we care? And if we do, what actions should we take? Since these questions about the planet affect all of us, then we must all share in the discussion. I divide into three groups those who sit next to me on planes and who engage in conversation. I characterize (and caricature) them as Mr. Smith, Mrs. Jones, and Dr. Brown.

Mr. Smith, Mrs. Jones, and Dr. Brown

Mr. Smith shows promise, because he asks if he can read my copy of *The Economist* if I'm through with it. We chat for a while. Mr. Smith cares about what is going on in the world; his company is a multinational one. Nonetheless, he does not think that environmental matters are a high priority. And he is inclined to think that people who study the environment are "greens" and that most tales of environmental damage are fabrications from a dangerous political fringe.

He is rather like *The Economist* on its worst days. It has excellent coverage of business news and by far the most competent and comprehensive international coverage in its class of weekly magazines. Indeed, it is likely to be the only coverage one can get of the situations in Afghanistan or Zaire. It has no environment section. And even though "greens" are ubiquitous and homogenous and "browns" do not exist, it is replete with accounts of the consequences of human impacts on the environment.

A tropical storm has caused landslides on deforested slopes killing 10,000 people. It will be impossible for the country to repay even the interest on its foreign debt. Somewhere in the Middle East, the agriculture ministers are threatening to go to war over water, and some countries are already firing shots at other countries' fishing vessels. A new hemorrhagic fever has broken out along some other deforestation front.

The Economist ended its second century of publication with a fat, special issue, crammed with superb essays about history, politics, religion, culture, science, exploration, the arts, but not the environment.[2] In its failure to see underlying causes and connections of the events it described, the magazine was a disappointment, but so are other forms of media. (A conspicuous exception in this case is CNN.)

As regular as clockwork, at the millennium's end, *The Economist* has yet another account of a tropical landslide where the body count may rise to half that of the American total in its war in Vietnam. Five years ago, when I started to write this book, I created a folder titled "landslides" into which I put news clippings from the Web. With sickening regularity I have entered stories of tens of thousands dying in mudslides throughout the world's once forest-rich mountainous, tropical countries. Assembling many of the same stories, Paul Ehrlich would write to me and ask: "Why did only one news story in 30 point to the root cause as the world's poor clearing forests because they had no other choice?"

Mr. Smith concedes my point about ubiquity of the unifying, environmental factors that underpin national and international news. I share with him this book's key statistics as summarized in the Prologue. "Because our impacts on the planet are so large they cannot help but play a major role in our lives, our politics, our history. And they will play an even bigger role," I tell him.

"Yes, but look at how much improvement in life span and quality of life our Western economies have brought!" he replies. Like the magazine, he believes that many of the world's problems can be fixed with participatory democracy and free-market economics. When there are major problems they are the result of a market or political failure. Thus, the global decline over the last 15 years in the amount of grain grown per person, a decline even more pronounced in Africa, is due largely to the fact that so many countries are political basket cases, he tells me. Mr. Smith and I have much to agree on. We begin to discuss details.

First is the squabble between the United States and Canada over salmon fishing, and next is the European Union's argument over who gets the Greenland turbot. I ask him how much he thinks it costs to buy an all-season boat to fish the waters off Greenland? And who made the loans? (The government, of course.) If he were a banker, would he have approved them? His look suggests that he would now be sweeping the bank's floor after hours had he done so.

A new dam in China is being opposed by the greens—who else? Mr. Smith is reminded of an obscure endangered fish, holding up progress. It was the snail darter and the Tellico dam in Tennessee a generation ago, I tell him. (It's the fish David Etnier finds in Chapter 11.) I also inform him that the dam builders' own economists knew that the dam consumed, not created, agricultural land, nor could it be justified on the basis of flood control or power generated. The flimsy justification of some politicians was that a new city of tens of thousands would be built along the lake's shores. No one else believed that it would happen. The dam was built and the city was not.

Reading on, we find the economic bottom line in China. Having dammed all the easy rivers, this one is going to be technically difficult and hugely expensive. Millions of people will be displaced. The health consequences to those who live nearby have not been estimated.

The article does not ask whether any fiscally responsible free-marketer would approve the massive squandering of money that this and many other water projects entailed, had it been private (and not the public's) money. International banks will not lend the project money unless their governments guarantee the loan. Like so many other schemes, this one will be an ecological disaster first and an economic one later.

The all-pervasive greens object to forest clearing. The fact that people live in those forests and some have died for expressing their belief in participatory democracy is probably overlooked. So, too, are the tax incentives and land claims that go to those who clear forests for cattle ranching.

"Greens versus jobs" is Mr. Smith's recollection of another headline. "Greens prevent massive squandering of public funds" is a headline only in my dreams. I would even accept "Fiscally responsible parties recommend against building the dam," ". . . against irrigating the dryland," ". . . against clearing the forest," ". . . against developing a new fishery," and so on, but those are as yet unrealized fantasies.

Mr. Smith and I may agree about how much better our lives are than our parents' and how we expect our children to be even better off. Our worlds are better. But whether the world itself is better is quite another matter.

Another flight, this one to Manaus, and on it, along with 40 others, Mrs. Jones is wearing a bright yellow T-shirt. All are in disgustingly good spirits given that it is the early hours of the morning, the flight is hours late, and the coffee is the usual packet to be poured into tepid water in a paper cup. Out come 40 bibles, on come 40 reading lights, and all possibility of sleep vanishes. In any case Mrs. Jones wants to save as many souls as she can and starts inquiring about mine. She asks about our destination; it's her first trip outside the United States, perhaps her first outside her state. We talk about the lives of those who eke out an existence in small villages tenuously connected to the outside world by dirt roads barely passable even when dry.

Her group dimly understands that they will have to save soils as well as souls. The small church they will build over the next two weeks will be a complete waste of effort if the small community impoverishes the land and abandons it. I tell her of another church's program in Haiti. On this ecologically devastated island, the church funds the installation of water tanks that retain the runoff, which then irrigates crops and trees. Their roots bind the soils, preventing the soils from washing away. In time, the trees will be large enough to harvest and their sale will fund more water tanks.

Mrs. Jones is like the great majority of Americans who attend a church, synagogue, mosque, or temple. Like them, she has a deep concern for human souls, one that she expresses in practical ways.

The conversation is going well, but I know the odds. The odds are that Mrs. Jones is like 50 percent of the undergraduates in the University of Tennessee biology class that I used to teach. She does not believe in evolution, despises those who teach it, and retains a special venom for those who teach it and who attend churches more accommodating than the one she attends. A few years ago, Tennessee came within a couple of votes of passing a two-paragraph law that would have fired teachers who taught evolution. I would have been one of them. I know that nothing I say to Mrs. Jones will convince her of the facts of evolution and the ages of the Earth and the universe. Why, then, do I continue my dialogue with her?

To allow this schism to continue means eschewing all discussion of the profound ethical consequences of humanity's current impact. If the past is any guide to the future, the hostilities will continue through this new century as they did the last. Unless we change the actions of our current population, let alone a future larger one, we will irreversibly destroy a large fraction of Earth's biodiversity within that interval.

I would not want Mrs. Jones on my local school board. But I do want to discuss with her the ethical, moral, and spiritual consequences of human actions in the environment and the loss of biodiversity. God's Creation, Mrs. Jones prefers to call it. I ask her, "Do you know that we may destroy a third of His Creation within this century?"

Dr. Brown is obvious as he boards the third flight. Bearded, he wears a T-shirt with "Earth does not belong to man, man belongs to Earth" emblazoned on the back, and drawings of bald eagles on the front. I guess correctly that he is on his way to a large meeting of professional ecologists. He tells me of his research on a bird in a national park. Some males "tweet" once, others twice as they hang upside down from a branch to attract mates. Single tweeters get fewer mates but live longer. He publishes his scientific papers in reputable ecology journals, ones I have read from cover to cover for a decade without ever seeing a hint that Earth has environmental problems.

Does he spend any time working on the science of such problems? "No." Has he helped the park's scientists or managers with the issues they face? "I've never met them." Has he shared the implications of his science with his congressional delegates? "That's for others." He hasn't even heard of the Union of Concerned Scientists and their frequent scholarly and measured evaluations of the major scientific issues facing American society. His disdain for such activities is matched by his supreme confidence that because he is an ecologist whatever he finds out must be important. I have absolutely nothing in common with Dr. Brown, as he will find out when I address his meeting after dinner that evening.

I have written this book in the hope that Mr. Smith and Mrs. Jones are among the readers. Realistically, I think that the Dr. Browns will predominate, so let me address him first.

Dear Dr. Brown . . .

It is the year 2001. Do you know where your planet is? The answer is an emphatic "no," and so I demand that you join with me in finding out. It is not simply that the ecologists of 2100 will be highly criticized if we fail. Our students will likely be unforgiving within our lifetimes, and their accusation could be a stinging one.

"You were the scientists, and in 2000 you couldn't answer even the simplest questions about the planet with any degree of precision. How many species were there? How much forest was there? How much land had been damaged by our use of it? How many rivers were dammed and channeled?" And so on down a short list of pressing global questions on terrestrial, freshwater, and marine systems.

Within a few decades these changes will be massive. Our inability to measure them accurately, compare their local rates, and understand what causes them may be more than an embarrassment. It may prove a fatal flaw in communicating with policy makers who too often use uncertainty as an excuse for inaction. Ignorance is not bliss. What might have happened if scientists had not measured the climate's temperature over the last century?

In answer to the students' accusation I start with species because they are central to all our activities. The estimates of the last decade span more than an order of magnitude—from a few million to a few tens of millions. That is a level of uncertainty greater than the estimates of the size of the universe. There are uncertainties in the global estimates of the loss of tropical forests. And there are large disparities in how much damage has been done to the world's drylands through desertification because there is so much disagreement over what the term *desertification* means. Similar concerns beset freshwater systems and the oceans.

I ask that we celebrate this new millennium by getting serious about ecology on a global scale. We need new skills, but crucially we must ask global questions and answer them at an appropriate level of accuracy. Yes, along the way we will make mistakes in their analysis. And we will correct them; it is what science does well. By 2010, we must be ready to answer the questions I have posed (and others I have implied) with enough accuracy to make us proud. By then I will be asking for even better accuracy in estimates of species, extinctions, and land use changes. But that is what scientists are supposed to do.

> My first recommendation, Dr. Brown, is that you help measure global parameters with a level of precision at least sufficient to detect a decade's worth of change.

Like a substantial fraction of scientists, you consider that you have no mandate to address the scientific needs of people who protect and manage species and their environments. This is simply inexcusable. I concede that one cannot always anticipate how scientific results will prove to be useful and important. We have many examples of obscure results that forever changed science. Darwin, remember, was a beetle collector. Nonetheless, numerous pressing scientific needs are associated with our massive impacts. We have an obligation to attend to them. If we don't do it, then who will?

"Oh, but I cannot get my papers published/get promoted/win the respect of my colleagues" [delete as appropriate] "if I do such applied work," Dr. Brown protests. The printable translation of my reply is "rubbish." The pages of *Nature* and *Science* contribute more to this book's material than any other sources. And many of this book's key figures have won the highest awards available in a discipline for which there is no Nobel Prize. The rewards for working on globally important problems are self-evident.

Where does one go to do such work? Tens, perhaps hundreds, of millions of people worldwide belong to groups that protect, manage, or restore forests, soils, drylands, rivers, fisheries, oceans, and biodiversity. Finding what needs to be done locally is easy. Nationally and globally, there are many organizations. Some are staffed mostly by scientists, so, naturally, scientific issues are of paramount concern in what they recommend.

Some have suggested that we tithe, giving a tenth of our time and effort to applied versus academic pursuits. That is a pathetically low amount, given the magnitude of the problems, but it is better than nothing.

> My second recommendation is that you become involved in the practical issues that beset our planet.

"What's the point," you ask, "if no one listens?" Yes, the media botch things, sometimes badly and sometimes deliberately. Reporters must please their readers, listeners, and viewers. They like to present both sides, however improbable one of them might be, because a fight makes good copy. For others, the journalistic tradition is not to evaluate the evidence but simply to present it and let the reader decide. While I have been slimed by reporters on occasion, the majority are keen to get the story right. The evidence of massive global impacts is not hard to present. Take the reporters to the scene, show them a photograph or a satellite image, explain the facts simply and with due attention to their uncertainties. These are all effective tactics. It simply takes time.

What about the politicians? You may be underwhelmed by your politician's voting record, but look at how he or she spends the day. In the United States, the House and Senate office building corridors are filled with teams of constituents and their paid helpers seeking more money than Willie Sutton ever dreamed of. Small groups, overnight bags in hand (in Washington, you must look as though you are from somewhere else), dash in and out of offices hoping for special legislation that will personally enrich them. Scientists who make similar visits want to inform and get nothing more than the member's vote. That can be effective, but it, too, takes time. Always remember, that every half hour you spend with your legislator is half an hour denied to that million-dollar-a-year lobbyist whose company wants to clear cut old growth forest on publicly owned land.

Does all this smack of "advocacy," the dirtiest eight-letter word in a scientist's vocabulary? We spend our lives educating our students and our colleagues. What part of our job description prohibits our extending that mission to the public and its political leaders?

My third recommendation: Communicate.

Dear Mr. Smith . . .

The economic impacts of human environmental actions are massive, ubiquitous, and unmistakable. Yes, Mr. Smith, I know the economists' self-effacing definition—an *economist* is someone who knows the price of everything and the value of nothing. I understand that market prices are easier objects of study, recorded abundantly in units—currencies—with known, if varying, calibrations. I know that you check your portfolio online three times a day and know its worth to the nearest cent. I sympathize when you tell me that values are more slippery, likely to vary widely from person to person and from generation to generation.

I understand, too, that prices reflect incremental or "marginal" costs. Two centuries ago Adam Smith noticed that diamonds traded for a higher price than did freshwater. (Perhaps today he would have compared cheaper gasoline with more expensive bottled water.) Yet the value of all freshwater is infinite: We could not survive without it. Across much of the planet, billions of people are worrying about where their children will find water. Prices poorly anticipate the consequences of humanity's growth, because they can change so quickly when the going gets tough.

These topics are too important to hide behind what is conceptually and computationally simple. There is far more to our human future than the detailed list of economic indicators at the back of my copy of *The Economist*. We must address the challenge of how to assess an environment's values and predict the future trends in those values.

In 1997, University of Maryland professor Bob Costanza organized a team of ecologists and economists at the National Center for Ecological Analysis and Synthesis in Santa Barbara, California, to address such questions. They took the flippant definition of economics to heart and estimated the value of all the world's ecosystems—essentially the value of everything. Their paper appeared in *Nature*, and its bottom line estimate—tens of trillions of dollars—was larger than the global sum of gross national products.[3] Angry criticisms appeared for months, and in 1998 *Nature* devoted a special segment to the debate.[4]

I hear your argument that the environment cannot be worth more money than all the money that we have to pay for everything else. Frankly, that is a smug, self-satisfied argument from those who wish to define a subject narrowly enough that they can study it rather than to encompass the more relevant but less tractable broader topic. Think about what you are worth and then suppose your house burns down. As everyone escapes, what few things do you rescue? The $100 cash or the irreplaceable photos of your late parents with your infant children that are valuable only to you and your family? Your worth is more than your bank account; so, too, is the planet's.

Some economists incorporate these views by asking how much the public would pay to keep something or by how much they feel they have lost when they do not. A real example of such calculations was when the court asked how deprived the public felt when the oil spill from the *Exxon Valdez* gummed up Alaska's scenically spectacular shoreline when it ran aground on Blyth's Reef in 1989. How much the American public felt that shoreline was worth—and not the actual market value of the subsequent decline in the region's tourism industry—was the amount of Exxon's fine.

Herein lies a vigorous debate. Some feel these calculations are far too speculative and uncertain, and so we should ignore them. That decision would imply that no one outside of Alaska or its tourists felt in any way deprived by Exxon's actions.

Other criticisms were about the technical details of the methods Costanza and his team used. Adding the most secure estimates still produces a global total of many trillions of dollars—and Costanza's team was unable to calculate many of the pieces needed for a complete sum. (And the team did not include values such as those illustrated by the *Exxon Valdez*.)

The global value of ecosystems services may be uncertain and difficult to calculate, but it is certainly huge and thus hugely inconvenient to ignore. Many of the same arguments apply to perverse subsidies. They, too, are on the scale of many trillions of dollars, yet their role in causing environmental harm—one known since Adam Smith—has been poorly investigated.

None of the critics proffered alternative estimates of the global total.[5] The most vigorous criticisms were overlain by the unmistakable expressions of pain that emanate from intellectuals when it is someone from a different field who sees something important.

How, then, should we proceed? I accept that my choice of *The Economist* as a convenient whipping boy is no solution. (Nor is that magazine in any way representative of economists as a profession. It's simply an excellent weekly magazine that caters to those who view the world through money-colored spectacles.) Beating up on economists doesn't solve anything either. Many now struggle to incorporate the value of the environment into economic assessments in a defensible manner. (Costanza was most harshly criticized by ecologists who thought his paper would harm that dialogue.)

A sensible response to Costanza comes from my Columbia University colleague Geoffrey Heal. In *Nature and the Marketplace: Capturing the Value of Ecosystem Services* he argues that what really matters is not heated debate on the total value of ecosystem services but practical solutions as to how we can capture those values.[6] Whatever the academic arguments are, we need mechanisms to protect what is valuable—and we need them now.

> Mr. Smith, my recommendation to you is that we must find ways to capture the value of the environment, encourage its sustainable use, and discourage activities that deplete it.

Yet others deem calculations of the value of nature to be irrelevant—a feeble, last gasp of economists to help them assess real values. Would either of us accept child labor just because willingness-to-pay estimates for its abolition were smaller than the money saved by paying children less than adults? Surely, there are overarching moral issues in providing a sustainable environment for future generations?

Dear Mrs. Jones . . .

It is no more excusable for you to overlook the massive environmental damage to the planet than it was for earlier generations to turn a blind eye to other acts of social injustice. A century ago, the "ethnic cleansing" was from the grasslands of the American West and the Argentine pampas. A century before that, it was the more temperate forests of eastern North America and Australia. Environmental destruction harms people directly today. In clearing tropical forests, we are also clearing them of indigenous peoples, destroying them as individuals and as cultures.

Environmental destruction also impoverishes the lives of future generations. Mrs. Jones, you don't think that having "dominion over every living thing" means the right for humanity to exterminate a third of God's Creation before we even know everything He created, do you?

I know you dislike Dr. Brown. The division between science and religion is because neither wants the other crossing the threshold of its temples. That makes the need for dialogue even more pressing.

> My recommendation to you, Mrs. Jones, is that the extent
> of human impact on the planet, its peoples, and its variety
> of life constitutes major ethical issues that demand a dialogue
> between science and religion.

The chances of this are better than you think. Dr. Brown may not go to your church, Mrs. Jones, but the odds are strongly in favor of his admitting to being "spiritual." It is a near certainty that you both share ethical concerns about environmental damage and the loss of biodiversity/creation. Surely neither of you has missed those key events in history where unlikely idealism and deep-rooted ethics triumphed over economic greed and self-centered hatred. The special issue of *The Economist* notices the decline of slavery, child labor, unsafe working conditions, racial discrimination, dictatorships, and does not mourn their passing. Dr. Brown may have his reservations about you, Mrs. Jones, but he must acknowledge the powerful legacy of ethical actions you bequeath to us all.

Dear Dr. Wilson and Dr. Moore . . .

Edward Wilson and Gordon Moore are quite real. The former, now retired from Harvard University, is one of the most influential and internationally honored biologists of our time. The latter, the cofounder of Intel, is the author of "Moore's Law"—the empirical doubling of computing power every 18 months or so, surely the most salient characterization of technology and industry at the start of the century.

They come to the podium of the meeting room at Cal Tech, in Pasadena, California, in August 2000. Both are softly spoken and share a quiet sense of humor. The podium is covered in laptop computers ready for the PowerPoint presentations to follow. Moore quips that he always enjoys seeing so many computers. They are the real representatives of the private sector and the academic community coming to discuss common concerns.

Together they are hosting a conference sponsored by the Center for Applied Biodiversity Science at Washington-based Conservation International, and The World Conservation Union headquartered in Gland, Switzerland. I'm there at Wilson and Moore's invitation as the organizer of the scientific program, having recruited 30 scientists to address the challenge they had posed: What would it take to defy nature's end?

As they reflected in their invitation letter:

Never have we been in such a position to create real change. The global conservation effort can be raised to the level the problem requires. We firmly believe that together we can make substantial and significant headway in our battle to save the world's precious and endangered biodiversity.[7]

"Be bold, be creative, and imagine that whatever resources needed are available," they admonished. A year of preparation and a week of intensive discussions later and we do have some answers.

Protect more of Earth's remaining natural ecosystems.

This is our obvious, overarching recommendation. Is this practical? If so, how should we proceed? Our deliberations involved a series of alternative ideas, some debated long into the night.

Is saving the remaining biodiversity still possible, or is its destruction unavoidable?

Human numbers are the root cause of the damage we do to the planet. Were our numbers and aspirations to grow continuously, natural areas would have no chance. In the long term, we must make the transition to sustainable use of natural resources. Yet for the present, it does seem possible to save life on Earth.

Croplands cover approximately 15 million km^2 and do not overlap extensively with the wilderness forests of the Amazon, Congo, New Guinea, or Siberia. The biodiversity hot spots have been more seriously impacted, but their total area is still relatively small. The major tragedy is that the loss of natural areas and biodiversity contributes pitifully little toward overall human welfare. As previous chapters explained, of the 7 million km^2 of cleared tropical humid forests, less than 2 million km^2 have remained productive cropland. The rest is usually abandoned, or contributes marginally to the global livestock production. About 10 percent of converted tropical forests are on steep mountain slopes with high rainfall where the consequences of deforestation to settlers are often tragic. The poor soils underlying some areas soon degrade and support neither agriculture nor biodiversity.

This does not mean that saving nature will be economically or politically simple. Those who clear forests from bad land do so because they are displaced, marginalized, and perceive no alternative. Nonetheless, the chances for saving nature would be spectacularly reduced were croplands to overlap more extensively with the areas holding most species. That no such broad overlap yet exists offers hope, though current expansion of agriculture, logging, and tree plantations could quickly dash this optimism.

In the oceans, fishing constitutes a major threat to biodiversity. Most major fisheries are overexploited. Establishing protected marine reserves should increase the supply of fish.

In short, the harm we inflict on biodiversity often stems from impacts on critically vulnerable areas that contribute relatively little to overall human welfare. In some forests, most drylands, and most fished ecosystems, our mismanagement diminishes both biodiversity and welfare.

We can have nature and eat too.

Given that the task of protecting nature is theoretically possible, is it economically possible or is this goal, though desirable, prohibitively expensive?

The conference delegates added up their estimates and argued passionately about them. Their efforts are wildly speculative, but the sums that emerge from these and other efforts[8] suggest that the totals, while large, undercut any argument that saving nature is economically infeasible. The total cost estimate is the same order of magnitude as the individual wealth of the world's richest citizens—hundreds of billions of dollars. That is one-hundredth of the estimates of Costanza and his colleagues for just the global value of the ecosystem services that nature provides each and every year.

This size of the total cost suggests a strategy of leveraging private sector involvement with funding from governments and international agencies. Such plans are starting to take shape. One example is the effort organized by the World Wildlife Fund, and advised by Larry Linden of Goldman Sachs, to help the Brazilian government protect the Amazon.[9]

Will protecting nature work or will the areas continue to be degraded?

The difficulty is in the details, for all conservation is local. Some countries have welcomed international conservation actions, recognizing the advantages of protecting forests (for example) *and* having much-needed funds. Other nations may not follow. Brazil may welcome the WWF initiative. But it also has elaborate plans to develop the Amazon, planning roads, channeling rivers, and expanding energy production at an estimated cost of

$45 billion to benefit its soy farmers.[10] Forest-rich countries top world lists of corrupt governments.[11] Cutting forests and otherwise depleting resources is too often a way to personal aggrandizement among some government officials. Furthermore, the pressures to convert forests are often external: The World Bank and the International Monetary Fund have indirectly encouraged governments to deplete their natural resources to pay off debt.[12] Even stable and corruption-free governments may view an infusion of foreign aid as imperialism in a twenty-first-century guise. Almost all the biodiversity hot spots were once European colonies (one is still a French territory), and many of them attained independence only after World War II.

Conservation groups may buy logging concessions to deny tropical forests to loggers, but loggers are not the only cause of deforestation. How good is a government's guarantee of a forest's security when its peoples need to cut wood to cook or heat, or to clear land for farming? Will the government return the concession fees to those whose livelihoods are most affected? Some countries are at war. In others, forests are far removed from whatever passes as national government.

The conference delegates' discussions of these questions quickly devolve into a myriad of individual experiences and idiosyncratic case histories that differ from country to country. There will not be a one-size-fits-all solution. So what we proposed was not a common solution, but a common process to engineer solutions. It was the establishment of centers to train and empower professionals in each country rich in biodiversity and wild areas. Presently, these capabilities are highly centralized in the United States and Europe, while key tropical areas have few conservation professionals. Some important countries have few biologists of any kind.

Should efforts concentrate on protection or on slowing the processes that harm nature?

If we are to avoid the breaking wave of extinction, then immediate protection and the immediate training of in-country professionals are necessary to implement and sustain that protection. Nonetheless, there was a healthy debate about how much effort should be allocated to actions that would lighten the burden of future generations of conservation professionals. Many of the delegates argued for immediate actions that could both stem the processes that harm nature and encourage those that protect it, with priority given to those that yield near-term results.

As Chapter 10 elaborated, perverse economic subsidies that degrade the environment emerge as a common problem across terrestrial, freshwater, and marine ecosystems. Likewise, perverse property rights deplete natural resources by giving ownership to those whose "economic use" often translates into short-term forest clear-cutting, transient crops or grazing, and longer-term land degradation. There are many important opportunities to save money *and* nature through targeted changes in policy.

> ## Do we know enough science to identify the areas to protect or is a major scientific effort necessary?

Yes, we need more scientific research—I made that recommendation to Dr. Brown earlier—but the total costs are well within the capacity of global scientific enterprise. The need for a greatly expanded research effort is particularly evident when one considers the poorly known linkages between biodiversity, ecosystems, their services, and people. Infectious diseases are entering human populations as our numbers increase and as we encroach on tropical forests and other pathogen reservoirs. Scarcity of safe drinking water already affects hundreds of millions of people and will affect 2 to 3 billion people over the next 25 years.[13]

Not Dreaming

In Wagner's opera *The Flying Dutchman*, the Dutchman moors his ship with its blood-red sails after the steersman falls asleep on the dock one evening. Dire things then happen for a couple of hours. Then the heroine, Senta, flings herself into the sea in typical Wagnerian fashion—for love, heroically, and accompanied by some great music. In some performances, the steersman wakes up and realizes that it has all been a bad dream. In other performances, members of the audience wake up with the same feeling.

I would like to tell you that after falling asleep during the Prologue in a bank of ferns on Maui, I awakened to find that this story about the world at the turn of the twenty-first century has been a bad dream, too. Wrong. The extent of our impacts is too obvious to miss, the consequences too serious to dismiss. At least in broad outline, *The World According to Pimm* is Mr. Smith's world, Mrs. Jones's world, and your world too. The science in this book adds the necessary quantification, perhaps pointing to details you previously overlooked. It is my fervent hope that you are now better informed.

The Pasadena meeting concluded with shared senses of concern, optimism, and urgency.[14] Concern because humanity's numbers are increasing. Individually, humanity will consume more, and many have unmet aspirations. If current trends of human impact continue, nature has no chance.

Across several human generations, a transition to sustainable use of natural resources is essential. Nothing precludes our conserving our irreplaceable biodiversity during that transition.

Our optimism stems from the realization that greatly increasing the areas protecting biodiversity represents a clear and achievable goal, one potentially attainable using funds raised in the private sector and leveraged through governments. A dialogue between hitherto mutually suspicious communities—Mr. Smith, Mrs. Jones, and Dr. Brown—has started, at this conference and elsewhere.

The urgency is driven by the pending loss of so much of our world's natural areas and biodiversity in the first half of this century if we do not act immediately. There is much more that we need to know, but we clearly know enough to act. Our world is a spectacularly beautiful, interesting, and diverse place. Only by attending to its problems will it remain so.

Notes

Comments on Data Sources

The Food and Agriculture Organization of the United Nations (FAO) at www.fao.org and the United Nations Environment Programme (UNEP) at www.unep.org are the sources of many statistics that appear throughout the book. A key database is the FAOSTAT Statistics Database that is available online. They do a brave job of compiling difficult-to-obtain data from many countries, some of which are probably less than enthusiastic about having to report, and many of which do not have the means to do so.

By far the most accessible source of these and other UN data are the *Yearbooks* from the World Resources Institute (WRI). I relied on two of these in particular: *World Resources: A Guide to the Global Environment 1996–97* and *World Resources: A Guide to the Global Environment 1998–99* (Oxford: Oxford University Press). Henceforth, I call these *WRI 1996–97* and *WRI 1998–99*, respectively. Each volume is in two parts. The first part of *WRI 1996–97* deals with the urban environment; the first part of *WRI 1998–99* deals with environmental change and human health. The second and larger parts of each volume are the data on global environmental trends.

The data themselves have problems. Often there is passionate reporting of far more significant figures than can be credible. Some countries do not report numbers, yet the *Yearbooks* blissfully total statistics by continent as if the numbers were exactly zero. For instance, the report's total number for charcoal consumption is likely to be too small because many charcoal-burning countries do not report their numbers.

These *Yearbooks* compound these problems with an extraordinary tendency to scramble numbers, miss numbers, have different numbers when they should be the same numbers and the same numbers for when they should be different numbers. Often the numbers in their tables do not add up to the totals they suggest. They manage to shrink the world's croplands by 2 million km^2 from the 1996–1997 to the 1998–1999 *Yearbooks*, but this turns out to be caused by an inability to correctly add the total areas in the latter book. The list of rapidly expanding tree plantations shows them growing at identical rates 10 years apart. Here (and elsewhere) the numbers are printed in volume after volume with no warning that they are the same numbers rehashed from one decade to the next. The tables almost always lack the documentation to tell the reader how the numbers were obtained and how old they were.

Comparing numbers between years (and especially yearbooks) can be an even more uncertain procedure. Deforestation statistics have increased dramatically, but that is more likely the result of changes in what is meant by that process than any underlying acceleration.

Reporting areas of natural forest and other environmentally interesting statistics may not be WRI's principal interest, but that merely reinforces the need to carefully check any numbers they produce.

Finally, there are the problems inherent in reporting data on a country-by-country basis. Large countries encompass many different ecosystems. India, for instance, is gaining forests according to FAO tables, but that is surely the large gain in plantations in some areas of that very large country being offset by a slightly smaller loss of natural forests in other smaller countries.

This said, the global accountant has few other choices but the FAO. I have found the tables to be a valuable resource once one understands their many limitations.

Conversions and Units

Not all of the numbers in these references are in the same units that I have used. The global change community uses units of Pg (petagram) or Gt (gigatons) of carbon—both are 1 billion tons. Throughout, I convert grams of carbon to grams of biomass by multiplying by 2.2. Some of the plant productivity data are in kilojoules; there are 4.19 joules per calorie. I use a conversion factor of 1 gram of biomass = 5 kilocalories justified in Chapter 2. Some data are given as grams/m² g · m⁻² in the convention I shall use. One g · m⁻² is equivalent to 1000 tons · km⁻². Other more specific conversions are provided in the corresponding notes.

Prologue

1. M. P. Moulton and S. L. Pimm, "The Introduced Hawaiian Avifauna: Biogeographical Evidence for Competition," *American Naturalist*, 121 (1983): 669–690; M. P. Moulton and S. L. Pimm, "The Extent of Competition in Shaping an Experimental Avifauna," in J. Diamond and T. J. Case, eds., *Community Ecology* (New York: Harper and Row, 1985).
2. J. E. Cohen, *How Many People Can the Earth Support?* (New York: W. W. Norton, 1995).

3. L. Peters and T. E. Lovejoy, eds., *Global Warming and Biological Diversity* (New Haven: Yale University Press, 1992).

Part 1 Green Forest, Yellow Desert

1. D. Adams, *The Hitchhiker's Guide to the Galaxy* (New York: Harmony Books, 1980).
2. P. M. Vitousek, P. R. Ehrlich, A. H. Ehrlich, and P. A. Matson, "Human Appropriation of the Products of Photosynthesis," *Bioscience*, 36 (1986): 368–373.
3. The border between Mongolia and the Russian district of Chita is visible from space, a consequence of the Soviets overgrazing their side. D. Sneath, "State Policy and Pasture Degradation in Inner Asia," *Science*, 281 (1998): 1147–1148.

CHAPTER 1 *Billions* of Tons of Green Stuff

1. S. L. Pimm and R. L. Kitching, "The Determinants of Food Chain Lengths," *Oikos*, 50 (1987): 302–307.
2. *WRI 1996–97* and *WRI 1998–99*.
3. A comparison of vegetation classifications is provided by World Conservation Monitoring Centre, *Global Biodiversity: Status of the Earth's Living Resources* (London, UK: Chapman & Hall, 1992).
4. J. S. Olson, J. A. Watts, and L. J. Allison, "Carbon in Live Vegetation of Major World Ecosystems." ORNL publication no. 5862. Oak Ridge National Laboratory, Oak Ridge, TN, 1983. The database is a CD-ROM and associated documentation: J. K. Kineman and M. A. Ohrenschall, *Global Ecosystems Database, Version 1.0* (Boulder, CO: U.S. Department of Commerce, National Oceanic and Atmospheric Administration, National Geophysical Data Center, 1992).
5. Since the Vitousek et al. study, there has been a concerted effort to compile the widely scattered estimates of primary production. Compilations by ORNL DAAC User Services Office, Oak Ridge National Laboratory, P.O. Box 2008, MS-6407, Oak Ridge, Tennessee 37831-6407 USA are now available online. (http://www-eosdis.ornl.gov/home.html). I have averaged these values to correspond to my best guesses as to which of Olson et al.'s classifications they refer. With the help of my colleague Robert Powell, I converted the Olson et al. data to a text file and then used its classification of each half degree latitude-longitude cells. This process forms the basis of the book's maps, though I have combined some areas to simplify the mapping. Finally, multiplying the areas derived for each of Olson's vegetation classes by the ORNL production estimates, I obtained an independent estimate of global terrestrial production. The following table summarizes these efforts.

Olson Class	Area (km²)	Production (tons · km⁻²)	Total Production (Gt)	Combined Areas	Corresponding Map
Wooded tundra	1,586,164	200	0.32		Map 2
Tundra	6,627,567	200	1.33	8,213,732	Map 2
Crops	15,420,721	1700	26.22		Map 3
Broadleaf temperate	6,036,951	1100	6.64		Map 4
Mixed forest	5,017,760	1100	5.52		Map 4
Forest and fields	5,932,905	1100	6.53	16,987,616	Map 4
Taiga	4,442,839	500	2.22		Map 5
Conifer forest	8,454,084	1000	8.45	12,896,923	Map 5
Tropical forest	11,422,873	1100	23.99		Map 6
Shrub trees thorns	3,902,775	500	1.95		Map 8
Woods, fields, savannas	10,883,186	1100	11.97	14,785,961	Map 8
Grasslands	21,951,805	530	11.63		Map 9
Shrub woodlands	3,411,897	370	1.26	25,363,702	Map 9
Deserts	9,356,578	200	1.87		Map 10
Semidesert	11,248,423	200	2.25	20,605,002	Map 10
Wetlands	2,440,086	1250	3.05		No map
Total	128,136,614		115.20		

Next, the numbers used by Vitousek et al.

Area	Area (10⁶ km²)	Production (tons · km⁻²)	Global Production (Gt)
Forests	31	1580	49
Dry woodlands, grasslands	37	1400	52
Deserts	30	100	3
Arctic and alpine areas	25	80	2
Agriculture	16	950	15
Swamps, marshes	6	1800	11
Cities and other human areas	2	200	0.4
Total	147		132

Suppose we compare the simplified schemes of Olson et al. and Vitousek et al. Both groups agree with the amount of land under cultivation. Olson et al. do not include Antarctica, Vitousek et al. lump it with all the "arctic and alpine" areas. For the nearly 16 million km² of ice cap that Vitousek et al. classify as tundra, their average productivity of 80 tons of biomass per square kilometer is surely too high. What happens if "zero tons" is a better number? This only makes a difference of 2 Gt—a little over 1 percent of the total.

The big differences are essentially a matter of how one classifies the different areas: Vitousek et al. have 30 million km² of desert, Olson et al. have only 21 million km² of desert and semidesert and another 3.4 million km² of shrublands. Clearly, one can dice and slice these categories in an endless number of ways (and botanists do!) but the bottom line estimates of production are not that different.

6. VEMAP Members 1995, "Vegetation/Ecosystem Modelling and Analysis Project: Comparing Biogeography and Biogeochemistry Models in a Continental-Scale Study of Terrestrial Responses to Climate Changes and CO² Doubling," *Global Biogeochemical Cycles,* 9 (1995):407–537; J. M. Melillo, "Global Change and Terrestrial Net Primary Production," *Nature,* 363 (1993):234–240. Climate modelers seem happy with an estimate of 60 Gt carbon for terrestrial production. This is exactly 132 Gt · tons of biomass (Vitousek et al.'s estimate) using 2.2:1 biomass to carbon conversion.

The most recent summary, employing satellite imagery to observe seasonal and regional variations on both land and the oceans is by C. B. Field, M. J. Behrenfeld, J. T. Randerson, and P. Falkowski, "Primary Production of the Biosphere: Integrating Terrestrial and Oceanic Components," *Science,* 281 (1998): 237–240. They estimate the annual productions (in Gt of carbon) of tropical rainforests, 17.8; broadleaf deciduous forests, 1.5; broadleaf and needle-leaf forests 3.1; needle-leaf evergreen forests 3.1; needle-leaf deciduous forest 1.4; savannas 16.8; perennial grasslands 2.4; broadleaf shrubs with bare soil 1.0; tundra 0.8; desert 0.5; cultivation 8.0. Converted to biomass, their total is 124 Gt of biomass for the land. For the oceans these authors estimate 107 Gt biomass.

There must be considerable variation in production from year to year as a consequence of events such as El Niño (which makes the Amazon drier than normal) and major volcanic eruptions (such as that of Mount Pinatubo, which caused global cooling). D. S. Schimel, B. H. Braswell, R. McKeown, D. S. Ojima, W. Parton, and W. Pulliam, "Climate and Nitrogen Controls on the Geography and Timescales of Terrestrial Biogeochemical Cycling," *Global Biogeochemical Cycles,* vol. 10 (1996): 677–692 argue that annual changes of 5 to 20 percent in production are possible during these climatic anomalies.

CHAPTER 2   What Earth Does for Us—And What We Do to Earth

1. I. V. S. Rombauer, I. S. Rombauer, and M, Rombauer Becker, *Joy of Cooking* (New York: MacMillan, 1975); I. S. Rombauer, M. Rombauer Becker, and E. Becker, *Joy of Cooking* (Running Press, 2000).
2. J. E. Cohen, *How Many People Can the Earth Support?* (New York: W. W. Norton, 1995).
3. Olson's estimates of calories. See ref. 4.
4. Food and Agriculture Organization of the United Nations (1993), Agrostat-PC, FAO, Rome, and *WRI Yearbooks.* All data are in wet weights and so have to be

converted. Using FAO data, Vitousek et al. calculated as follows. The world total production is about 1.7 Gt of grain, of which 1.1 Gt (65 percent) is for humans, the rest for livestock. This is wet weight: It represents 0.85 Gt dry weight. We also consume 0.3 Gt of nongrains, for a total of 1.15 Gt. Livestock consumption of grains is 35 percent of 1.7 Gt = 0.59 Gt, wet weight, or 0.46 Gt, dry weight.

5. *The New York Times*, February 27, 1998, p. A10.
6. F. B. Morrison, assisted by E. B. Morrison and S. H. Morrison, *Feeds and Feeding: A Handbook for the Student and Stockman* (Ithaca, NY: Morrison Publishing Company, 1956).
7. These are my estimates of animal consumption, using FAO data on numbers of animals and Morrison data on consumption.

Animal	Total (10^9)	Daily Consumption (g)	Year Total (Gt)
Humans (for reference)	5.0*	500	0.91
Cows	1.28	1500	0.70
Sheep and goats	1.78	500	0.32
Pigs	0.86	1000	0.32
Horses, mules, donkeys	0.12	4000	0.17
Buffaloes and camels	0.16	4000	0.23
Total animal			1.74

*The number used by Vitousek et al.; it has increased by 1 billion since then.

8. The FAO/WRI data are given in cubic meters, so we require another conversion factor. A cubic meter of water weighs exactly 1 metric ton. (That is by design and not by coincidence.) Wood is less dense than water—that's why it floats, of course—and its average density is about 0.6, according to Vitousek et al. Thus 1 m^3 of wood weighs 0.6 metric tons. From these data, the total industrial consumption of wood is 1.6 billion m^3. Some is sawn into planks, and some made into panels and paper. Vitousek et al., using different sources of data, came up with a slightly higher total: 1.0 Gt for just boreal and temperate forests, with tropical areas contributing another 0.3 Gt.

   WRI 1996–1997 estimates a total of 1.856 billion m^3 of fuel and charcoal production per year for 1991 to 1993. (These convert to biomass; see ref. 2.7.) There are other problems with this data beyond the reporting of more significant figures than can be credible. Some countries do not report numbers, yet the *Yearbook* totals statistics by continent as if these numbers were exactly zero. As a consequence, the report's total number for charcoal consumption is too small.

   To get data on per capita consumption I take data on charcoal consumption from p. 220 divided by population data on p. 190 in *WRI 1996–1997*.
9. Results come from comparing recent *WRI Yearbooks* to data in Vitousek et al.
10. *WRI Yearbooks* estimate about 15 million km^2 of croplands, the same number as for Olson et al. (ref. 1.5). The latter source includes urban areas for a total of 15 million km^2. Vitousek et al. estimate these at about 2 million km^2 but then

Olson includes a category of mixed fields and woods (about 6 million km²). The "fields" portion includes crops as well as grazing land. Splitting this 6 million into three equal shares, 2 for croplands (which balances the 2 subtracted from the 15 for urban areas), 2 for grazing, and 2 for forests is only a rough guess, but it is one that is concordant with other estimates of the three pieces.

11. United Nations Population Division, Interpolated National Populations 1950–2025 (The 1994 Revision); on diskette (New York: UN, 1995). Also reproduced in *WRI 1996–1997*.

12. M. Reisner, *Cadillac Desert: the American West and Its Disappearing Water* (New York: Penguin, 1993).

CHAPTER 3   Not the Forest Primeval

1. Which ecosystems does agriculture consume? I obtained the totals by comparing two data sets on the Global Ecosystems Database CD-ROM (version 1.0): see ref. 1.4. The first is the Olson et al. classification of croplands. This tells us what are crops, but not what the land was originally. The second classification is the Holdridge classification of the planet's original vegetation. It does this by a temperature by rainfall table. Average annual temperatures range from 3°C to 30°C in five divisions (boreal, cool temperate, warm temperate, subtropical, and tropical). Rainfall is classified into seven divisions, from 0.125 m · yr⁻¹ to 8 m · yr⁻¹. These divisions roughly correspond to desert, desert bushes, steppe, and forests from "dry" to "rain." The warmer the area, the less plant growth a given rainfall will support. Half a meter of rain in boreal areas supports a moist forest, but in warmer areas the same amount will support only a steppe or thorn forest.

My colleague Robert Powell created a data file consisting of the sites identified in half degree latitude and longitude blocks considered to be croplands by Olson et al. He then merged this file with the Holdridge classification of what

	Desert (km²)	Desert Bush (km²)	Steppe to Forest (km²)	Forest (km²)	Forest (km²)	Forest (km²)	Forest (km²)	Total Area (km²)
Rainfall (m/yr)*	0.125	0.25	0.5	1.0	2.0	4.0	8.0	
Polar (1.5°C)**								
Polar (3°C)	5053	12999	31587	28251				77890
Boreal (6°C)	9281	106582	399006	166522	51048			732439
Cool-temperate (12°C)	69271	94646	2043835 (steppe)	2925182	272989	19753		5425676
Warm-temperate (18°C)	52870	60089	471672	1489206	529999	18520		2622356
Subtropical (24°C)	235137	220484	349898	1441037	1765154	299809	17209	4328729
Tropical (30°C)	89728	59835	99337	436485	1059668	379897	27219	2152169
Total area	461340	554635	3395335	6486683	3678858	717979	44429	**15339259**

*Rainfall is per year.

**Temperatures shown are annual averages.

these sites would have been originally. For a tiny fraction of the sites, this process suggests that crops grow in areas classified as "polar." These are likely to be sites where crops grow in valleys amid extensive mountainous areas. These numbers likely underestimate the bite taken by agriculture from forests, for there Olson et al. have another category of forests and fields, some of which will be croplands (see ref. 2.8).

2. See ref. 2.8.
3. See refs. 1.5 and 2.8. The total for the three areas mapped is 17 million km^2 but 2 million km^2 have been allocated to crops and another 2 million km^2 to grazing lands within the forests.
4. Ecological histories of eastern North America include G. G. Whitney, *From Coastal Wilderness to Fruited Plain: A History of Environmental Change in Temperate North America from 1500 to the Present* (Cambridge, UK: Cambridge University Press, 1994). W. Cronon, *Changes in the Land: Indians, Colonists, and the Ecology of New England* (New York: Hill and Wang, 1991).
5. F. Jennings, *The Invasion of America: Indians, Colonization, and the Cant of Conquest* (Chapel Hill: University of North Carolina Press, 1975). The role of disease in the colonization of the New World and generally why we conquered them (and not the other way around) is explained by Jared Diamond in his prize-winning overview of world history: J. M. Diamond, *Guns, Germs, and Steel* (New York: W. W. Norton and Company, 1999).
6. Some of the quantitative estimates are listed in J. Loewen, *Lies My Teacher Told Me: Everything Your American History Textbook Got Wrong* (New York: The New Press, 1996), chap. 3, note 43.
7. Whitney, Chapter 10, pp. 227ff.
8. The fictitious trek westward is a description of the figures in S. L. Pimm and R. Askins, "Forest Losses Predict Bird Extinctions in Eastern North America," *Proceedings of the National Academy of Sciences (USA)*, 92 (1995): 9343–9347.
9. L. I. Wilder, *Little House in the Big Woods* and *Little House on the Prairie* (New York: Combined Edition, Harper Trophy, 1989). Yes, I read these books to my girls to get them to sleep. They are wonderful stories. They are also ecological histories of what Americans did to forests, prairies, and to the native people and animals that lived there. The books are worth another read with this perspective.
10. Whitney, p. 177.
11. Ibid., p. 191.
12. Ref. 3.8.
13. Ref. 1.5.
14. A. Shvidenko and S. Nilsson, "What Do We Know about the Siberian Forests?" *Ambio*, 23 (1994): 396–404.
15. A. Rosencrantz and A. Scott, "Siberia's Threatened Forests," *Nature*, 355 (1992): 293–294.
16. P. Saich, W. G. Rees, and M. Borgeaud, "Detecting Pollution Damage to Forests in the Kola Peninsula using the ERS SAR," *Remote Sensing of Environment*, 75 (2001): 22–28. They estimate that 400 km^2 have been "severely damaged" by the smelter; they report that damage is visual at distances up to 60 km from the smelter and that the damaged area is expanding at approximately 2 km per year.

17. Whitney, p. 209.
18. Ibid., Chapter 9.
19. Ibid., p. 198.
20. P. R. Ehrlich and J. P. Holdren, "Impact of Population Growth," *Science,* 171 (1971): 1212–1217.
21. See ref. 3.14.
22. Whitney, p. 257.
23. Data from Pimm and Askins (ref. 3.8) and with apologies to my friend Eric Chivian, who drives this route in a Saab and whose ancestors are not those described.
24. Whitney, p. 165.
25. This account is based on a briefing update on forest health prepared as a special project of Union of Concerned Scientists, Global Resources Department. Staff scientist Darren Goetze and UCS consultant Dr. Doug Boucher were major contributors. The estimate of fire damage was updated as of September 12, 1996, according to the National Interagency Fire Center in Boise, Idaho. Key references in this report included W. W. Covington and M. M. Moore, "Post-Settlement Changes in Natural Fire Regimes and Forest Structure: Ecological Restoration of Old-Growth Ponderosa Pine Forests," in R. N. Sampson and D. L. Adams, eds., *Assessing Forest Ecosystem Health in the Inland West* (New York: Food Products Press, 1994). Dominick A. DellaSalla,, David M. Olson, and Saundra L. Crane, "Ecosystem Management and Biodiversity Conservation: Applications to Inland Pacific Northwest Forests," in D. Baumgartner and R. Everett, eds., *Proceedings of a Workshop on Ecosystem Management in Western Interior Forests* (Pullman, WA: Washington State University Cooperative Extension Unit, 1995); R. N. Sampson, D. L. Adams, S. Hamilton, S. P. Mealey, R. Steele, and D. Van De Graaff, *Assessing Forest Ecosystem Health in the Inland West. Forest Policy Center, American Forests* (Washington, DC, 1994); R. W. Gorte, *Forest Fires and Forest Health,* Congressional Research Service Report No. 95-511 ENR (Washington, DC, 1995); Garland N. Mason, Kurt W. Gottschalk, and James S. Hadfield, "Effects of Timber Management Practices on Insects and Diseases," in R. M. Burns, technical compiler, *The Scientific Basis for Silvicultural and Management Decisions in the National Forest* System (Washington, DC: USDA-Forest Service General Technical Report WO-55, 1989).
26. See ref. 3.14.
27. Food and Agriculture Organization of the United Nations, *State of the World's Forests,* 1999.
28. R. Carrere and L. Lohmann, *Pulping the South* (London: Zed Books, 1996).
29. FAO, *Forest Resources Assessment 1990.*
30. Vitousek et al. for their estimate of tree plantations quote by G. L. Ajtay, P. Ketner, and P. Duvigneaud, "Terrestrial Primary Production and Phytomass," pp. 129–182, in B. Bolin, E. T. Degens, S. Kempe, P. Ketner, eds., *The Global Carbon Cycle* (New York: John Wiley and Sons, 1979). Their estimate is that plantations produce 2.6 Gt of biomass per year. Vitousek et al. also write that plantations represent about 4.8 percent of the total forest area. It is surprisingly difficult to confirm this.

The WRI's *1996/1997 Yearbook* is a dismal source of such information. It reproduces the numbers in the FAO document *Forest Resources Assessment 1990*. This provides numbers on a country-by-country basis, but there are important gaps in the records. There are no data for North America and Europe, for example. For the 26 million km² for which they do present data, the percentage of plantations is about 2 percent, or roughly 0.50 million km².

Sandra Postel and Lori Heise (1988) compile another incomplete set of numbers on plantations in their Worldwatch Paper 83: *Reforesting the Earth*. They do have some numbers for the areas that the FAO document omits and other numbers that allow comparisons. A comparison of these two sources estimates plantation cover at a minimum of 1.2 million km², with many countries not reporting or underreporting.

There are some obvious problems with these numbers. The various estimates for China, for example, differ by a factor of two—an amount more than the area estimated for North America. The 1990 FAO report does not include Europe or North America and the estimates here are clearly a problem. The estimate of 130,000 km² for plantations in Europe needs to be compared with the FAO estimate of more than 1 million km² of forest and woodland (*World Resources Yearbook 1996–1997*, table 9.1). Does this mean that most of Europe's forests are wild ones? Clearly not! Almost all of the forests are managed to greater or lesser degrees. The need for good global statistics on plantations is evident.

CHAPTER 4   When Vegetation Rioted and Big Trees Were King

1. Amazon fire data comes from a report by Stephan Schwartzman of the Environmental Defense Fund, downloaded January 2001; http://www.edf.org/pubs/Reports/AmazonBurning/.
2. H. M. Stanley, *Through the Dark Continent* (Mineola, NY: Dover Publications, reprinted 1988).
3. 1924, the U.S. Department of Agriculture *Yearbook*.
4. N. Myers, *The Primary Source: Tropical Forests and Our Future* (New York: W. W. Norton and Company, 1992, rev. ed.).
5. Food and Agriculture Organization of the United Nations, *State of the World's Forests*, 1999.
6. Ref. 1.5.
7. N. M. Collins, J. A. Sayer, and T. C. Whitmore, eds., *The Conservation Atlas of Tropical Forests: Asia and the Pacific* (New York: Simon and Schuster, 1991); C. S. Harcourt and J. A. Sayer, eds., *The Conservation Atlas of Tropical Forests: The Americas* (New York: Simon and Schuster, 1996); J. A. Sayer, C. S. Harcourt, and N. M. Collins, eds., *The Conservation Atlas of Tropical Forests: Africa* (New York: Simon and Schuster, 1992). The *Conservation Atlas* considers all tropical forests for the Americas and "closed canopy broadleaf forests" for the other two volumes.
8. Comparisons of FAO data are not a good source for the increasing rate of deforestation. They more likely represent changing ideas on what deforestation is supposed to be imposed by outside sources critical of their original low

estimates. In the Amazon, INPE (see ref. 4.12) have the most credible estimates. There is much year-to-year variation, but the overall trend is an increasing one.

9. N. Myers and J. Kent, *Environmental Exodus: an Emergent Crisis in the Global Arena* (Washington, DC: Climate Institute, 1995).

10. D. Skole and C. J. Tucker, "Tropical Deforestation and Habitat Fragmentation in the Amazon: Satellite Data from 1978 to 1988," *Science*, 260 (1993): 1905–1910.

11. *WRI 1990–1991*.

12. Brazil's Instituto Nacional de Pesquisas Espaciais (São Paulo: São José dos Campos) produces reports almost annually on deforestation containing tables on rates of deforestation by state and for the entire Brazilian Amazon. The most recent (2000) was *Monitoramento da Foresta Amazonica Brasileria por Satélite 1998–1999* and is available online at www.inpe.br. The annual rates in square kilometers averaged over the one or more years are 21,130 (1977 to 1988); 17,860 (1988 to 1989); 13,810 (1989 to 1990); 11,130 (1990 to 1991); 13,785 (1991 to 1992); 14,896 (1992 to 1994); 29,059 (1994 to 1995); 18,161 (1995 to 1996); 13,227 (1996 to 1997); and 17,383 (1997 to 1998).

13. See ref. 4.12.

14. D. C. Nepstead, A. Verssimo, A. Alencar, C. Nobre, E. Lima, P. Lefebvre, P. Schlesinger, C. Potter, P. Moutinho, E. Mendoza, M. Cochrane, and V. Brooks, "Large-Scale Impoverishment of Amazonian Forests by Logging and Fires," *Nature*, 398 (1999): 505–508.

15. J. C. Goldammer, "Forests of Fire," *Science*, 284 (1999): 178–183.

16. M. A. Cochrane, A. Alencar, M. D. Schulze, C. M. Souza, Jr., D. C. Nepstad, P. Lefebvre, and E. A. Davidson, "Positive Feedbacks in the Fire Dynamic of Closed Canopy Tropical Forests," *Science*, 284 (1999): 1832–1835.

17. T. M. Brooks, S. L. Pimm, and N. J. Collar, "Deforestation Predicts the Number of Threatened Birds in Insular Southeast Asia," *Conservation Biology*, 11 (1997): 382–384.

18. Food and Agriculture Organization of the United Nations, *State of the World's Forests*, 1999.

19. See ref. 4.18.

20. *The Economist*, May 11, 1996, pp. 14–17.

21. B. Hahn et al., "AIDS as a Zoonosis: Scientific and Public Health Implications," *Science*, 287 (2000): 607–614.

22. Olson (see ref. 1.5) has almost 11 million km^2 of "woods, fields, and savannas" (see Map 8). Comparison of this map and Map 6 (tropical moist forests) shows substantial overlap in the Amazon and Congo Basins, and Central America: These are areas of converted moist forest. Elsewhere, the areas shown in Map 8 are natural savannas (the llanos, in northern South America, for example).

23. A. Grainger, "Estimating Areas of Degraded Tropical Lands Requiring Replenishment of Forest Cover," *The International Tree Crops Journal*, 5 (1988): 31–36. Grainger estimates that 20.8 million km^2 of the tropics are degraded. Much of this area is drylands, the subject of Chapter 5. The total, however, includes 4.2 million km^2 of dry and montane forest, 1.4 million km^2 of tropical rainforests,

and 2 million km^2 of "forest fallows" in the humid tropics that are "in need of protected regeneration."

24. *WRI 1990–1991*.
25. D. G. Burnett, *Masters of All They Surveyed: Exploration, Geography, and a British El Dorado* (Chicago: The University of Chicago Press, 2000).

CHAPTER 5   Peace and Quiet and Good Earth

1. Nothing beats reading Cook in the original! J. C. Beaglehole, ed., *The Journals of Captain James Cook on His Voyages of Discovery* (Cambridge, UK: Cambridge University Press: Published for the Hakluyt Society. Vol. I, 1955, Vol. II, 1962, Vol. III, pts. 1 and 2, 1967).
2. J. M. Diamond, "Easter's End," *Discover* (August 1995), pp. 62–69.
3. J. M. Diamond, "Paradises Lost," *Discover* (November 1997), pp. 68–78.
4. H. C. Darby, *Domesday England* (Cambridge, UK: Cambridge University Press, 1977).
5. M. A. F. Kassas, "Ecology and Management of Desertification," in *Earth '88: Changing Geographic Perspectives* (Washington, DC: National Geographic Society, 1988).
6. United Nations Environment Programme (UNEP), *Status of Desertification and Implementation of the United Nations Plan of Action to Combat Desertification, Nairobi, Kenya*, 1977.
7. See ref. 1.5: the total areas of Maps 8, 9, and 10.
8. See ref. 6.
9. H. E. Dregne, "Desertification of Arid Lands," in F. El-Baz and M. H. A. Hassan, eds., *Physics of Desertification* (Dordrecht, The Netherlands: Martinus, Nijhoff, 1986); H. E. Dregne, *Desertification of Arid Lands* (New York: Hardwood Academic, 1983); H. E. Dregne and N. T. Chou, *Global Desertification Dimensions and Costs in Degradation and Restoration of Arid Lands* (Lubbock, TX: Texas Tech University, 1992).
10. D. Pimentel, C. Harvey, P. Resosudarmo, K. Sinclair, D. Kurtz, M. McNair, S. Crist, L. Shpritz, L. Futton, R. Saffouri, and R. Blair, "Environmental and Economic Costs of Soil Erosion and Conservation Benefits," *Science*, 267 (1995): 1117; R. Lal and B. A. Stewart, *Soil Degradation* (New York: Springer-Verlag, 1990).
11. See ref. 7.5. Postel et al. estimate that 2.5 million km^2 of irrigated croplands require 2000 Gt of water, an equivalent rainfall of 80 cm.
12. E. Kessler, D. Alexander, and J. Rarick, "Duststorms from the U.S. High Plains in Late Winter 1977—Search for Cause and Implications," *Proceedings of the Oklahoma Academy of Science*, 58 (1978): 116–128.
13. See ref. 2.8.
14. The FAO numbers are as follows: Great Britain has 11,688,000 cows plus 29,375,000 sheep on 11,112,000 ha; New Zealand has 8,456,000 cows plus 51,730,000 sheep on 13,577,000 ha; Swaziland has 660,000 cows plus 449,000 sheep and goats on 1,070,000 ha; Afghanistan has 1,500,000 cows, plus 16,350,000 sheep and goats on 30,000,000 ha; Saudia Arabia has 202,000 cows plus 11,091,000 sheep and goats on 120,000,000 ha.
15. See ref. 1.5.

16. E. A. Pearce and G. Smith, *World Weather Guide* (New York: Times Books, 1990).
17. G. W. Thomas, T. W. Box, and J. L. Schuster, "The Brush Problem in Texas," *Brush Control Research Progress Report* (1968).
18. J. Hastings and R. Turner, *The Changing Mile* (Tucson, AZ: The University of Arizona Press, 1965). This is one of the most visually compelling studies of ecological history.
19. See ref. 5.8.
20. C. J. Tucker and S. N. Nicholson, "Variations in the Size of the Sahara Desert from 1980 to 1997," *Ambio*, 28 (1999): 587–591.
21. A case history discussed in ref. 5.6.
22. Technically, the text discusses the coefficient of variation, the standard deviation of summer rainfall over 30 years divided by the 30-year mean.
23. S. L. Pimm and A. Redfearn, "The Variability of Animal Populations," *Nature*, 334 (1988): 613–614. The variability (and not just the range) of animal population numbers increases over the interval during which they are calculated. So, too, do climatic variables as the following reference notes.
24. M. Schroeder, *Fractals, Chaos, Power Laws: Minutes from an Infinite Paradise* (New York: W. H. Freeman, 1992).
25. J. Raban, *Bad Land* (London, UK: Picador Press, 1996).
26. D. Worster, "The Dirty Thirties," *Great Plains Quarterly*, 6 (1996): 107–116.
27. See ref. 4.8.
28. G. C. Daily, "Restoring Value to the World's Degraded Land," *Science*, 269 (1995): 350.

CHAPTER 6 "Man Eats Planet! Two-Fifths Already Gone!"

1. R. S. DeFries, C. B. Field, I. Fung, J. Collatz, and L. Bounoua. "Combining Satellite Data and Biogeochemical Models to Estimate Global Effects of Human-Induced Land Cover Change on Carbon Emissions and Primary Productivity," *Global Biogeochemical Cycles*, 13 (1999): 803–815.
2. P. M. Vitousek, H. A. Mooney, J. Lubchenco, and J. M. Melilo, "Human Domination of Earth's Ecosystems," *Science*, 277(1997): 494–499
3. Atmospheric carbon dioxide may be measured with exquisite sensitivity, but where it comes from and where it goes to are not easily measured. The former is simple chemistry; the latter involves all the complex and uncertain ecological calculations described in Chapters 1 through 5 for land and Chapters 8 through 10 for oceans. Recent estimates put the contribution of deforestation to increasing carbon dioxide at 1.5 to 2 Gt of carbon per year, which equals 3.3 to 4.4 Gt of dry weight biomass. I. Fung, "Variable Carbon Sinks," *Science*, 290 (2000): 1313; P. Bousquet, P. Peylin, P. Ciais, C. Le Quéré, P. Friedlingstein, and P. P. Tans, "Regional Changes in Carbon Dioxide Fluxes of Land and Oceans Since 1980," *Science*, 290 (2000): 1342–1346. These estimates are only a third to a half of the Vitousek et al. estimate of 9 Gt, who quoted W. Seiler and P. J. Crutzen, "Estimates of Gross and Net Fluxes of Carbon between the Biosphere and the Atmosphere from Biomass Burning," *Climate Change*, 2 (1980): 207–247. However, some fraction of the forest that is burned will remain as charred logs and ash and some of the carbon will be the fine particulate matter in smoke. Forest clearing also destroys a lot of biomass in addition to what is

burned. So a measurement of what the atmosphere gains in carbon dioxide will still underestimate the biomass losses from human actions.

CHAPTER 7   Water, Water Everywhere?

1. N. Myers, "South Africa's Growing Pains," *South Africa's Growing Pains. People and the Planet*, 5, no. 3 (1996): 31.
2. United Nations Population Division, Interpolated National Populations 1950–2025, 1994 revision. On diskette (New York: United Nations, 1995). Also reproduced in *WRI 1996–1997*.
3. S. Postel, *Last Oasis: Facing Water Scarcity* (London, UK: W. W. Norton and Company, 1992). She quotes J. R. Starr, "Nature's Own Agenda: A War for Water in the Mideast." *The Washington Post*, March 3, 1991. See also F. Pearce, "Wells of Conflict on the West Bank," *New Scientist*, June 1, 1991, as examples of the language that attends conflicts over water. See also ref. 7.7.
4. See ref. 7.2.
5. WRI 1996–1997.
6. See ref. 7.3.
7. *The Golan Heights Important Facts* is a document soliciting donations from English-speaking readers for the permanent occupation of this part of Syria by 14,000 Israelis. Produced by the Golan Residents Committee, June 1995.
8. *WRI 1996–1997*.
9. See ref. 7.3.
10. P. Gleick, "Basic Water Requirements for Human Activities: Meeting Basic Needs," *Water International*, vol. 21 (1996): 83–92.
11. S. L. Postel, G. C. Daily, and P. R. Ehrlich, "Human Appropriation of Renewable Freshwater," *Science*, vol. 271 (1996): 785–788.
12. See ref. 5.15.
13. It was close to being available as the manuscript was completed; C. Nilsson, personal communication, January 2001.
14. M. Dynesius and C. Nilsson, "Fragmentation and Flow Regulation of River Systems in the Northern Third of the World," *Science*, 266 (1994): 753–762. Their unit measurement of flow rate is $1 \text{ m}^3 \cdot \text{s}^{-1}$. That is equal to 1 ton/s, or $3600 \text{ s} \times 24 \text{ h} \times 365 \text{ days} = 31.5$ million tons per year.
15. D. M. McArthy and C. W. Voigtlander, eds., *The First Fifty Years: Changed Land, Changed Lives* (Tennessee Valley Authority, 1983).
16. E. A. Norse, ed., *Global Marine Biological Diversity* (Washington, DC: Island Press, 1993), pp. 149ff.
17. Quoted from ABCnews.com, July 22, 1997.
18. Ref. 7.3, p. 123.
19. P. P. Micklin, "Desiccation of the Aral Sea: A Water Management Disaster in the Soviet Union," *Science*, 241 (1988): 1170–1176.
20. The comparison of the two images of the Small Aral Sea in 1962 and 1987 was at edcwww.cr.usgs.gov/dclass. A major Aral Web site is: www.geology.sdsu.edu/geology/facilities/carre/index.html.
21. The source of Butakoff's map is www.lib.utexas.edu/Libs/PCL/Map_collection/historical/Aral_1853.jpg.
22. See ref. 7.14.

Part 2   Blue Ocean, Green Sea

1. D. J. Roy, R. W. Old, and B. E. Wynne, eds., *Bioscience and Society* (Chichester, UK: John Wiley, 1991).
2. D. Pauly and V. Christensen, "Primary Production Required to Sustain Global Fisheries," *Nature,* 374 (1995): 255–257.
3. L. Speer, L. Lauck, E. K. Pikitch, S. Boa, L. Dropkin, and V. Spruill, *Roe to Ruin: The Decline of Caspian Sea Sturgeon and the Road to Recovery* (Washington, DC: Caviar Emptor Campaign, 2000); R. DeSalle and V. J. Birstein, "PCR Identification of Black Caviar," *Nature,* 381 (1996): 197–198.

CHAPTER 8   On the Hero's Platform

1. A. Longhurst, S. Sathyendranath, T. Platt, and C. Caverhill, "An Estimate of Global Primary Production in the Ocean from Satellite Radiometer Data," *Journal of Plankton Research,* 17 (1995): 1245–1271. Their estimates of production were in grams of carbon. The lowest values of production are in the North Atlantic (75 mg of biomass equivalent produced/$m^{-2} \cdot d^{-1}$), and the highest, the Benghula current (4225 mg/$m^{-2} \cdot d^{-1}$).
2. D. Pauly and V. Christensen, "Primary Production Required to Sustain Global Fisheries," *Nature,* 374 (1995): 255–257. The production numbers quoted by the next reference (in grams of carbon/$m^{-2} \cdot yr^{-1}$) are 103 for open ocean, 973 for upwellings, 310 for tropical and temperate shelves, 890 for coastal and reef systems, and 290 for rivers and lakes. The principal source for their estimates is G. G. N. De Vooys, pp. 259–292, in B. Bolin, E. T. Degens, S. Kempe, and P. Ketner, eds., *The Global Carbon Cycle* (New York: Wiley, 1979).
3. Longhurst et al. (ref. 8.1) suggest a global estimate of 53 Gt carbon, equivalent to 117 Gt biomass. Field et al. (ref. 2.6) estimate total marine production to be 107 Gt of biomass.
4. Nanoplankton are smaller than 20/1000ths and picoplankton are smaller than 2/1000ths mm.
5. F. P. Chavez and R. T. Barber, "An Estimate of New Production in the Equatorial Pacific," *Deep Sea Research,* 34 (1987): 1229–1243.
6. A summary of Kendra Daly's work, which discusses some of the factors that affect the measurements of phytoplankton production, can be found in K. L. Daly and W. O. Smith, Jr., "Physical-Biological Interactions Influencing Marine Plankton Production," *Annual Reviews of Ecology and Systematics,* 24 (1993): 555–585.
7. D. M. Karl, R. Leteller, D. Hebel, L. Tupas, J. Dore, J. Christian, and C. Winn, "Ecosystem Changes in the North Pacific Subtropical Gyre Attributed to the 1991–92 El Niño," *Nature,* 373 (1995): 230–234.
8. W. O. Smith, Jr., L. A. Codispoti, D. M. Nelson, T. Manley, E. J. Buskey, J. Niebauer, and G. F. Cota, "Importance of *Phaocystis* Blooms in the High Latitude Ocean Carbon Cycle," *Nature,* 352 (1991): 514–516. P. A. Wheeler, M. Gosselin, E. Sherr, D. Thibault, D. L. Kirchman, R. Benner, and T. E. Whitledge, "Active Cycling of Carbon in the Central Arctic Ocean," *Nature,* 380 (1996): 697–699. K. R. Arrigo and C. R. McClain, "Spring Phytoplankton Production in the Western Ross Sea," *Science,* 266 (1994): 261–263.

9. The Coastal Zone Color Scanner flew into space aboard the Nimbus-7 satellite in 1978. The Web site ishttp://daac.gsfc.nasa.gov/data/dataset/CZCS/index.html. The Coastal Zone project has now been replaced with the SeaWiFs, the details of which are at http://neptune.gsfc.nasa.gov/seawifs.html.

10. To compute production, one needs a mathematical model that relates surface color to what goes on below the near surface. This combination of detailed satellite images over time and space and depth modeling constitute the recipe of Longhurst et al. (see ref. 8.1). Their model uses over 26,000 samples of depth profiles of chlorophyll. Even so, their estimates of production in the equatorial Pacific are low compared to the recent shipboard estimates I mentioned earlier.

11. The largest discrepancies in estimating marine production come from the complications of calculating open ocean productivity. In the main body of text, I used Pauly and Christensen's figure of 225 tons/$km^{-2} \cdot yr^{-1}$. In the units common to oceanographers, this is 103 g $C/m^{-2} \cdot yr^{-1}$.

Longhurst et al. calculate an annual production of 113 g $C/m^{-2} \cdot yr^{-1}$ for the Pacific Equatorial Divergence. Chavez and Barber (whose figures they do not quote) come up with values of at least 176 g $C/m^{-2} \cdot yr^{-1}$ for the same region—nearly 60 percent larger. The uncertainties about the open oceans increase if we go to the North and South Pacific gyres, the least productive parts of the ocean. Here Longhurst et al. estimate 59 and 87 g $C/m^{-2} \cdot yr^{-1}$, respectively. Karl et al. (*Nature*, 1995) have yearly rates equivalent to 140 g $C/m^{-2} \cdot yr^{-1}$ even before the El Niño event—and readings are 40 percent higher during an El Niño event. Data are from 100 km N of Oahu, 22.45 N, 158 W.

CHAPTER 9   Lots of Good Fish in the Sea

1. Huxley's Inaugural Address is from M. Foster and E. R. Lankester, eds., *The Scientific Memoirs of Thomas Henry Huxley* (London: Macmillan, 1898).

2. *Vanity Fair*, January 18, 1871.

3. FAO. *The State of the World Fisheries and Aquaculture* (Rome: FAO, 1995).

4. Primary productivity is often reported in dry weights (the measure I used in Chapters 7 and 8), but sometimes in calories, or joules (both measures of energy) or sometimes in weights of carbon. Pauly and Christensen used a 9:1 ratio of wet weight to carbon, and I've multiplied by 2.2 to get back to dry weight. So dry weight is the wet weight divided by 4.1 (= 9/2.2).

5. See ref. 9.3.

6. C. Safina, *Song for the Blue Ocean* (New York: Henry Holt and Company, 1997).

7. L. Watling and E. A. Norse, "Disturbance of the Seabed by Mobile Fishing Fear: A Comparison to Forest Clearcutting," *Conservation Biology*, 12 (1998): 1180–1197.

8. See ref. 9.6.

9. A guide to what should be eaten is M. Lee, ed., *The Seafood Lovers Almanac* (New York: National Audubon Society, 2000). It also shows what the fish we eat look like, how they are caught, and (of course!) how to cook them.

10. FishBase is at www.fishbase.org. By March 2001, it contained information on 25,241 species, 71,000 synonyms, 108,000 common names, 28,000 pictures, 21,000 references, and was the product of 550 collaborators.

11. FAO, *A Global Assessment of Fisheries by Catch and Discards.* FAO Fisheries Technical paper no. 339 (Rome: FAO, 1994); M. A. Hall, "On Bycatches," *Reviews in Fish Biology and Fisheries,* 6 (1996): 319–352.

12. See ref. 7.16, p. 94.

13. Hall, ref. 9.11.

14. D. Pauly, V. Christensen, J. Dalsgaard, R. Froese, and F. Torres, Jr., "Fishing Down Marine Food Webs," *Science,* 279 (1998): 860–863.

15. Details of pizza toppings are from *Food and Wine,* September 1996, p. 14.

16. A highly readable guide to the oceans is E. A. Norse, ed., *Global Marine Biological Diversity* (Washington, DC: Island Press, 1993).

17. The fishery statistics are from FAO, *The State of the World Fisheries and Aquaculture* (Rome: FAO, 1995). See also *A Global Assessment of Fisheries Bycatch and Discards,* FAO Fisheries Technical Paper no. 339 (Rome: FAO, 1994).

18. S. Nicol, "Antarctic Krill," in W. A. Nierenberg, ed., *Encyclopedia of Environmental Biology,* vol. 2 (San Diego: Academic Press, 1995). Details on the catch were obtained from an Environmental News Service report, December 14, 1999, quoting data from the Commission for the Conservation of Antarctic Marine Living Resources.

19. See ref. 9.17.

CHAPTER 10   The Wisdom to Use Nature's Resources

1. The accounts of Crosbie's visits to St. John's and to the coastal villages are from John DeMont's articles in the July 6, 1992, issue of *Maclean's,* the weekly Canadian news magazine, and from news items in the three succeeding issues.

2. M. Kurlansky, *Cod: A Biography of the Fish That Changed the World* (New York: Walker and Company, 1997).

3. A. Moorehead, *The Fatal Impact* (New York: Harper and Row, 1987) is a compelling and beautifully written account of the cultural and ecological slaughter that followed Cook's mapping of the Pacific.

4. G. Daws, *Shoal of Time: A History of the Hawaiian Islands* (Honolulu, HI: University of Hawai'i Press, 1968).

5. *Encyclopaedia Britannica,* vol 19, pp 291–293.

6. P. O'Brian, *The Far Side of the World* (London, UK: W. W. Norton and Company, Limited, 1984); this is the tenth book in the Aubrey and Maturin series of historical novels and is an entertaining and detailed account of the wars and whalers of the period.

7. See ref. 2.1.

8. See ref. 9.17.

9. National Marine Fisheries Service, 1999. Report to Congress: Status of the fisheries of the United States. Available at www.nmfs.gov.

10. J. M. Casey and R. A. Myers, "Near Extinction of a Large, Widely Distributed Fish," *Science,* 281 (1998): 690–692.

11. G. Hardin, "The Tragedy of the Commons," *Science,* 162 (1968): 1243–1248.

12. B. J. McCay and A. C. Finlayson, "The Political Ecology of Crisis and Institutional Change: The Case of the Northern Cod." Paper presented to the annual meeting

of the America Anthropological Association, November 15–19, 1995. That paper, downloaded from the Web, anticipated a published paper by Finlayson, "Linking Social and Ecological Systems: Institutional Learning for Resilience," in a volume to be edited by F. Berkes and C. Folke, Sweden, Beijer Institute.

13. C. Safina, "The World's Most Imperiled Fish," *Scientific American*, 273 (1995): 46–53.
14. C. S. Baker, G. M. Lento, F. Cipriano, M. L. Dalebout, S. R. Palumbi. M. Goto, and S. Ohsumi, "Scientific Whaling: Source of Illegal Products for Market?" *Science*, 290 (2000): 1695–1696. C. S. Baker and S. R. Palumbi, "Which Whales Are Hunted? A Molecular Genetic Approach to Monitoring Whaling," *Science*, 265 (1994): 1538–1539.
15. See ref. 7.3.
16. C. W. Clark, "Bioeconomics," in R. M. May, ed., *Theoretical Ecology* (Sunderland, MA: Sinauer, 1981, 2d ed.).
17. See ref. 9.6.
18. N. Myers, "Lifting the Veil on Perverse Subsidies," *Nature*, 392 (1998): 327; N. Myers and J. Kent, *Perverse Subsidies: How Tax Dollars Can Undercut Both the Environment and the Economy* (Washington, DC: Island Press, 2001). FAO estimated in 1993 that fisheries worldwide spent $54 billion annually to subsidize a catch worth $70 billion at market. More recent estimates confirm fishing revenues at $70 to $80 billion, but place the subsidy at between $20 to $30 billion. Details can be found in M. Milazzo, *Subsidies in World Fisheries* (Washington, DC: World Bank, 1998).
19. M. Reisner, *Cadillac Desert: the American West and Its Disappearing Water* (New York: Penguin, 1993).
20. N. Myers and J. Kent, *Perverse Subsidies: How Tax Dollars Can Undercut Both the Environment and the Economy* (Washington, DC: Island Press, 2001).
21. See ref. 10.18.
22. A. Smith, *An Inquiry into the Nature and Causes of the Wealth of Nations* (1776). The discussion of bounties is in Book 4, chap. 5, pp. 242–262. Herrings are the subject of pp. 249–250.
23. See Part 12, Protection and Preservation of the Marine Environment, United Nations Conference on the Law of the Seas III. On the Web at www.globelaw.com/LawSeas.
24. See ref. 10.10.

Part 3   The Variety of Life

1. R. M. May, "An Inordinate Fondness for Ants," *Nature*, 341 (1989): 386–387.
2. M. Williamson, "High Table Tales," *Nature*, vol. 341 (1989): 695; May's untitled reply follows.

CHAPTER 11   An Inordinate Fondness for Beetles

1. D. A. Etnier and W. C. Starnes, *The Fishes of Tennessee* (Knoxville, TN: The University of Tennessee Press, 1993).

2. Nigel Stork, personal communication, December 2000.
3. *Oreophasis derbianus, Aulacorhynchus derbianus, Psittacula derbiana, Eriocnemis derbyi, Orthotomus derbianus and Serinus rothschildi, Bangsia rothschildi, Cypseloides rothschildi, Astrapia rothschildi, Leucopsar rothschildi.*
4. E. Mayr, *Animal Species and Evolution* (Cambridge, MA: Harvard University Press, 1963).
5. B. D. Patterson, "The Species Alias Problem," *Nature*, 380(1996):589.
6. R. M. May, "The Dimensions of Life on Earth," in P. H. Raven and T. Williams, *Nature and Human Society: The Quest for a Sustainable World*, Washington, D.C.: National Academic Press, 1997, pp. 30–45.
7. R. G. Gillespie, S. R. Palumbi, and H. B. Croom, "Multiple Origins of a Spider Radiation in Hawaii," *Proceedings of the National Academy of Sciences*, 91 (1994): 2290–2294; R. G. Gillespie, "Hawaiian Spiders of the Genus *Tetragnatha:* III. *T. acuta* Clade," *Journal of Arachnology*, 22 (1994): 161–168; R. G. Gillespie, "Hawaiian Spiders of the Genus *Tetragnatha:* II. Species from Natural Areas of Windward East Maui," *Journal of Arachnology*, 20 (1992): 1–17; R. G. Gillespie, "Hawaiian Spiders of the Genus *Tetragnatha:* I. Spiny Leg Clade," *Journal of Arachnology*, 19 (1991): 174–209.
8. D. L. Hawksworth, "The Fungal Dimension of Biodiversity: Magnitude, Significance, and Conservation," *Mycological Research*, 95 (1991): 441–456.
9. N. E. Stork, "Insect Diversity: Facts, Fiction and Speculation," *Biological Journal of the Linnean Society of London*, vol. 35 (1988): 321–337.
10. T. L. Erwin, "Tropical Forests: Their Richness in Coleoptera and Other Arthropod Species," *Coleopterists Bulletin*, 36 (1982): 74–75.

CHAPTER 12   . . . Abode His Destined Hour, and Went His Way

1. S. L. Pimm, testimony to the Committee on the Environment, The Reauthorization of the Endangered Species Act, July 13, 1995.
2. See ref. 4.23.
3. J. Klicka and R. M. Zink, "The Importance of Recent Ice Ages in Speciation: A Failed Paradigm," *Science*, 277 (1999): 1666–1669.
4. J. T. Tanner, *The Ivory-Billed Woodpecker.* National Audubon Society Research Report 1 (New York: National Audubon Society, 1942).
5. S. L. Olson and H. F. James, "Descriptions of Thirty-Two New Species of Birds from the Hawaiian Islands: Part I. Non-Passeriformes," *Ornithological Monographs*, 45 (1991), Washington, DC: American Ornithologists' Union; H. F. James and S. L. Olson, "Descriptions of Thirty-Two New Species of Birds from the Hawaiian Islands: Part II. Passeriformes," *Ornithological Monographs*, 46 (1991). Washington, DC: American Ornithologists' Union.
6. S. L. Pimm, M. P. Moulton, and J. Justice, "Bird Extinctions in the Central Pacific," *Philosophical Transactions of the Royal Society*, 344 (1994): 27–33.
7. D. W. Steadman, "Prehistoric Extinctions of Pacific Island Birds: Biodiversity Meets Zooarchaeology," *Science*, 267 (1995): 1123–1131.
8. J. Curnutt and S. L. Pimm, "How Many Bird Species in Hawai'i and the Central Pacific Before First Contact?," in S. Conant and C. van Riper, III, eds., *Studies in Avian Biology*, (Lawrence, KS: Allen Press, 2001).

9. N. Collar, M. Crosby, and A. Stattersfield, *Birds to Watch* 2 (Washington, DC: Smithsonian Institution Press, 1994).

10. Serious scholars would expect me to quote all 3000 pages of C. Kidd, D. Williamson, and L. Collins, eds., *Debrett's Peerage and Baronetage 2000* (London, UK: St. Martin's Press, 2001). Personally, I quite like R. Buskin, *The Complete Idiot's Guide to British Royalty* (London, UK: Macmillan, 1997). And, no I don't expect to get a knighthood after publishing the paragraphs that follow.

11. S. L. Pimm, H. L. Jones, and J. M. Diamond, "On the Risk of Extinction," *American Naturalist*, 132 (1988): 757–785.

12. S. L. Pimm, *The Balance of Nature? Ecological Issues in the Conservation of Species and Communities* (Chicago, IL: University of Chicago Press, 1991).

13. S. L. Pimm, "The Snake that Ate Guam," *Trends in Ecology and Evolution*, 2 (1987): 293–295.

14. World Conservation Monitoring Centre, *Global Biodiversity: Status of the Earth's Living Resources* (London, UK: Chapman & Hall, 1992).

15. Cowling, R. M., ed., *The Ecology of Fynbos* (Cape Town, South Africa: Oxford University Press, 1992).

16. J. D. Williams, M. L. Warren, Jr., K. S. Cummings, J. L. Harris, and R. J. Neves, "Conservation Status of Freshwater Mussels of the United States and Canada," *Fisheries*, 18 (1992): 6–22.

17. R. R. Miller, J. D. Williams, and J. E. Williams, "Extinctions of North American Fishes During the Past Century," *Fisheries*, 14 (1989): 22–38.

18. TNC, Priorities for Conservation: *1996 Annual Report Card for U.S. Plant and Animal Species.* (Arlington, VA: The Nature Conservancy, 1996).

19. G. F. Maine, *Rubaiyat of Omar Khayam and Other Writings by Edward Fitzgerald* (London, UK: Collins, 1953).

CHAPTER 13   Nature's Eggs in Few Baskets

1. R. D. Keynes, ed., *The Beagle Record: Selections from the Original Pictorial Records and Written Accounts of the Voyage of the* H.M.S. Beagle (Cambridge, UK: Cambridge University Press, 1979).

2. C. G. Sibley and J. E. Ahlquist, *Phylogeny and Classification of the Birds* (New Haven, CT: Yale University Press, 1990).

3. L. L. Manne, T. M. Brooks, and S. L. Pimm, "Relative Risk of Extinction of Passerine Birds on Continents and Islands," *Nature*, 399 (1999): 258–261.

4. R. K. Colwell and D. C. Lees, "The Mid-Domain Effect: Geometric Constraints on the Geography of Species Richness," *Trends in Ecology & Evolution*, 15 (2000): 70–76.

5. D. Quammen, *Song of the Dodo: Island Biogeography in an Age of Extinctions* (New York: Scribner, 1996; Touchstone Books, 1997).

6. N. Myers, R. A. Mittermeier, C. G. Mittermeier, G. A. B. da Fonseca, and J. Kent, "Biodiversity Hotspots for Conservation Priorities," *Nature*, 403 (2000): 853–858.

7. Melanie Stiassny, personal communication, January 2001.

8. D. McAllister, F. W. Schueler, C. M. Roberts, and J. P. Hawkins, "Mapping and GIS Analysis of the Global Distribution of Coral Reef Fishes on an Equal-Area

Grid," in R. Miller, ed., *Advances in Mapping the Diversity of Nature* (London, UK: Chapman & Hall, 1994), pp. 155–175.

9. See ref. 3.8.

10. T. Brooks and A. Balmford, "Atlantic Forest Extinctions," *Nature*, 380 (1996): 115; T. M. Brooks, S. L. Pimm, and N. J. Collar, "Deforestation Predicts the Number of Threatened Birds in Insular Southeast Asia," *Conservation Biology*, 11 (1997): 382–384; T. M. Brooks, S. L. Pimm, V. Kapos, and C. Ravilious, "Threat from Deforestation to Montane and Lowland Birds and Mammals in Insular Southeast Asia," *Journal of Animal Ecology*, 68 (1999): 1061–1078.

EPILOGUE

1. See ref. pro.2.

2. *The Economist*, December 31, 1999.

3. S. L. Pimm, "The Value of Everything," *Nature*, 387 (1997): 231–232.

4. E. Masood and L. Garwin, "Costing the Earth: When Ecology Meets Economics," *Nature*, 395 (1998): 426–427.

5. See ref. epi.4. I have talked to many economists about the Constanza et al. effort since having written the accompanying News and Views in *Nature*. I am seen as an apologist for this effort. Far from it: I am critical of some of their numbers. What I find interesting is the spectacular failure of the economics profession to come up with an order-of-magnitude estimate of values that, almost certainly, are orders of magnitude larger than (say) the relatively piddling quantities they normally consider. "Better to quantify something small accurately and understand its dynamics in detail than go after large, fuzzy, and important numbers" seems to be the argument. Many economists have assumed Costanza made the mistake of calculating the total value of ecosystem services. (In which case, their number would have been a serious underestimate of infinity.) This is not so: They clearly calculated the annual, marginal values, as stated in the text and the extensive footnotes.

6. G. Heal, in *Nature and the Marketplace: Capturing the Value of Ecosystem Services* (Washington, DC: Island Press, 2000).

7. Details of the conference are available at www.defyingnaturesend.org.

8. A. N. James, K. J. Gaston, and A. Balmford, "Balancing the Earth's Accounts," *Nature*, 401 (1999): 323–324.

9. L. Alderman, "Special Report: Investing in Philanthropy, Saving The Rain Forest," *Barron's*, December 18, 2000.

10. G. Carvalho, A. C. Barros, P. Moutinho, and D. Nepstead. "Sensitive Development Could Protect Amazonia Instead of Destroying It," *Nature*, 409 (2001): 131. Avança Brasil. Os Custos Ambientais para a Amazonia (Belém, Brazil: Instituto de Pesquisa Ambiental da Amazonia, 2000).

11. A survey undertaken by Transparency International. Details from http://www. transparency.de/documents/newsletter/2000.3/third.html.

12. N. Sizer and D. Plouvier, "Increased Investment and Trade by Transnational Logging Companies in Africa, the Caribbean, and the Pacific," joint report for World Wide Fund for Nature-Belgium, World Resources Institute, and WWF-International.

13. See ref. 7.8.

14. S L. Pimm, M. Ayres, A. Balmford, G. Branch, K. Brandon, T. Brooks, R. Busta-
     mante, R. Costanza, R. Cowling, L. M. Curran, A. Dobson, S. Farber, G. Fon-
     seca, C. Gascon, R. Kitching, J. McNeely, T. Lovejoy, R. Mittermeier, N. Myers,
     J. A. Patz, B. Raffle, D. Rapport, P. Raven, C. Roberts, J. P. Rodríguez, A. Rylands,
     C. Tucker, C. Samper, M. L. J. Stiassny, C. Safina, J. Supriatna, D. Wall and D.
     Wilcove, "Can We Defy Nature's End?," *Science*, 233 (2001): 2207–2208.

A Guide to Areas and Weights

METRIC UNITS	COMPARABLE MEASURES
Area	
1 square meter (m^2)	ca. 1 square yard (yd^2)
1 hectare	100 m by 100 m or 2 U.S. football fields
1 kilometer (km^2)	ca. $\frac{1}{10}$ of a square mile (mi^2); 100 hectares
100 km^2	twice the size of Bermuda
1,000 km^2	Martinique, Hong Kong
10,000 km^2	Jamaica
100,000 km^2	Kentucky, the former East Germany, Iceland
1,000,000 km^2	Egypt or Bolivia
10,000,000 km^2	China, Europe, or the United States
Weights	
1 kilogram (kg)	ca. 2 pounds
1 metric ton (1000 kg)	almost exactly a nonmetric ton; the weight of 1 m^3 of water

Index

United States (*Cont.*)
  degradation of Southwest, 89–91
  increasing forest area of, 49
  overfishing in, 168–169
  prairie destruction in, 46
  temperate forests in, 37–41
  temperate rain forests in, 47
Upwellings, 135, 139, 142, 147, 157–158
Urban areas, 32, 100
U.S. Army Corps of Engineers, 174

V
Value of environment, 241–243
Viruses, 68–69
Vitousek, Peter, 10, 18, 27, 103

W
Wallace, Alfred Russel, 225–226
Wars, freshwater as cause of, 109–111
Washington, George, 43
Waste:
  from biomass consumption, 31–32, 102
  bycatch from, 158–160
  from lumbering, 44
Water(s)
  amounts used, 111–112
  clockwise/counterclockwise flow of, 135
  coastal, 139, 141, 142

Waters (*Cont.*)
  international, 169
  rainwater vs. irrigation, 83
  See also Freshwater; Oceans
Watling, Les, 152
The Wealth of Nations (Adam Smith), 175
West Africa, tropical forest in, 59
Wetlands:
  freshwater, 122–123
  production of biomass in, 22
Whales, 136–137
Whaling, 127, 166–168
Whitney, Gordon, 38, 43, 44
Wilder, Laura Ingalls, 40
Williamson, Mark, 181
Wilson, Edward, 229, 244–245
World Bank, 17, 66, 247
World Resources Institute, 17, 51, 61, 62, 73
World Resources Yearbook, 17–18, 51, 61, 73
World Wildlife Fund, 246
Worster, Donald, 96

Y
Yearbook (USDA), 55

Z
Zink, Bob, 205
Zooplankton, 154–156, 162

About the Author

STUART PIMM, PH.D., is Professor of Conservation Biology at the Center for Environmental Research and Conservation at Columbia University in New York. He was the recipient of a Pew Scholarship for Conservation and the Environment (1993) and an Aldo Leopold Leadership Fellowship (1999).

Pimm works at the interface between conservation and policy, and has been called upon to testify before both the House and Senate committees on the re-authorization of the Endangered Species Act. He has been energetic in working for the restoration of the Florida Everglades. Pimm serves on the Science Advisory Council to Conservation International, on the Committee on Research and Exploration of the National Geographic Society, and on the board of the Union of Concerned Scientists.

Pimm maintains active field research programs in the Everglades, Madagascar, the Atlantic coast forests of Brazil, Central America, and South Africa. He and his research team survey birds and mammals in the field and analyze satellite images to determine what remains of their habitats in order to set effective priorities for conservation action. He is the author of more than 150 scientific papers, as well as three books and numerous popular articles and book reviews in such publications as *New Scientist*, *The Sciences*, *Nature*, and *Science*.

He maintains an energetic international lecturing schedule, and appears on television regularly, most recently on such shows as ABC News with Peter Jennings, CNN , the Discovery Channel, TV Asahi (Japan), and ABC (Australia). Ed Wilson (of Harvard) and Pimm were the subjects of a 1996 BBC Horizon (in the United States, PBS's Nova) feature entitled Nature's Numbers. Pimm's activities are routinely reported in the national press including the *New York Times* and the *Miami Herald*.